custom
The Concept of

La Triennale di Milano Foundation

LA TRIENNALE DI MILANO

PERMANENT COLLECTION OF ITALIAN DESIGN

custom-built

The Concept of Unique in Italian Design

Edited by Silvana Annicchiarico

CHARTA

Printed in Italy
ISBN 88-8158-433-6
Edizioni della Triennale

La Triennale di Milano Foundation
viale Alemagna 6
20121 Milan
tel. +39-0272434.1
fax +39-0289010693
www.triennale.it

Catalogue edited by
Silvana Annicchiarico
Design
Daniele Mastrapasqua
Iconographic research and bibliography collaborators
Giorgio Galleani
Translation
Judith Mundell

Edizioni Charta

Via della Moscova 27
20121 Milano
tel +39-026598098/026598200
fax +39-026598577
email: edcharta@tin.it
www.chartaartbooks.it

Permanent Collection of Italian Design

fuori serie
pezzi unici, prototipi e prodotti su commissione nell'archeologia del design italiano

Palazzo della Triennale
Viale Alemagna 6, Milan
April 8 – July 13, 2003

Curators
Silvana Annicchiarico
Giuseppe Raboni
Marco Zanuso jr
Exhibition Design
Giuseppe Raboni, Marco Zanuso jr
Lighting Consultant
Piero Castiglioni
Graphic Design
AR.CH.IT Architetti Associati
Object Research Collaborators
Luca Cipelletti, Alessandro Pedretti,
Rosanna Monzini

Installation 1960 snc, Mediterranea srl
Wiring Marzoratimpianti
Illuminated Ceiling Italian Ceiling per Barrisol
Exhibition Graphics Cip srl
Shipping Zust Ambrosetti
Insurance Mansutti spa

For their collaboration in realizing the exhibition installation, we would like to thank

iGuzzini MARZORATI RONCHETTI

Contents

The Age of Furniture

Italian Furnishings, Materials and Interior Design of the Thirties and Forties

Arturo Dell'Acqua Bellavitis
Professor, Politecnico di Milano

The exhibition *Custom-Built* is part of a project whose aim is to enrich La Triennale di Milano's Permanent Collection of Italian Design. In view of a possible museum of design that would be organized thematically and exhibited in rotation at the main venue (while others might be exhibited abroad or in other Italian museums), broadening the temporal horizon of the collection to include the thirties, which so energetically welcomed the dawning of furnishings production, is of particular interest.
The exhibition seems especially significant on the eightieth birthday of La Triennale, where exhibitions at the end of the twenties voiced lively debates on the quality of furnishings. Compared to the two previous editions, the 3rd Monza Biennial of 1927 was characterized by increased selectivity. In Marangoni's

words, "While the first two exhibitions involved a mass mobilization of Italian manufacturers, the third was intended to show the increasingly rigorous choice that marks from one exhibition to the next an effective progress of artists and a continuous elevation of products."[1]

At the 1927 Biennial rustic tradition and furnishing decorations imitating previous styles yielded to compact lines and enclosed volumes. German magazines like *Moderne Bauformen*, *Deutsche Kunst und Dekoration* and *Die Kunst* contributed to the spread of a taste for simplification, and for the strong, square fashion that would subsequently be known as novecento. Neo-classical architects, creators of "moderate" works, imitated these forms of expression. Their distaste for decoration and their bolder ideas were confined to exploiting the simplification of form and the clarity of concepts that classical style could provide them. Gio Ponti and Emilio Lancia were two of these architects who shunned tradition, "which meant taking a somewhat childish stance against Carducci, D'Annunzio and Umberto-style rhetoric," and took part in the Monza Biennial. Their complete habitats, characterized by unpretentious neo-classicism, were the fruits of a furnishings firm, Domus Nova, part of Rinascente, which was intent on reproducing the analogous experience of Parisian department stores in Italy.

With the 5th Triennial there was a positive leap in quality compared to previous editions. The Triennial increasingly became a venue for international exhibitors in both the architectural and figurative fields. In fact, in 1933 the International Biennial of Decorative Arts legally became the "International Triennial of Decorative, Industrial and Modern Arts and Architecture," with the

aim of shifting the focus away from craft products, which had in previous exhibitions prevailed over the aesthetics of the industrial object. The exhibition was moved from Villa Reale in Monza to the Palazzo dell'Arte, which was expressly built in Parco Sempione in Milan by Giovanni Muzio. This move further indicated a desire to be linked more closely with industry, a link that would prove in the next decade, however, to be truer in theory than in practice.
These initiatives were grafted onto a cultural climate now characterized by the "modern tradition," expressed by more sensitive circles who were deeply committed to the struggle for the renewal of the arts (such as Group 7, made up of Figini, Pollini, Terragni, Frette, Rava, Larco, Libera, and Miar in Rome). At the fourth edition, Figini and Pollini presented the "Casa Elettrica," built in Sempione Park. In addition to being an important approach to the issue of economical housing, it was also an exhaustive statement on Italian rationalism.
Foreign participation, however, especially middle-European, was gradually increasing, and introduced innovations through solo exhibitions (Mies, Gropius, Le Corbusier, Aalto and others, as well as whole foreign sections, especially at the 4th and 5th Triennials). In the furnishings sector, Aalto's furniture of 1928, presented at the 6th Triennial, constituted a veritable experiment in which the Finnish maestro, with the help of craftsman Otto Korhonen, made the first "flexible" armchair. The fundamental concept that characterized Aalto's furniture and architecture was the creation of "a smooth, ribbon-like shape," as opposed to a rigid structure of joints. Naturally the "forwardness" of this exhibition did not escape the bureaucracy of the time (let's not forget that we are on the thresh-

old of the thirties), which attempted at least in part to "curb its effects" with craft industry demands, patriotic claims and academic theorization.
The *Custom-Built* exhibits plainly reveal the coexistence of different forms of research into furnishings and, by association, interior design. Alongside experiments involving new idioms that looked to the other side of the Alps with the use of metal or new techniques for bending wood, we also find practices bearing a resemblance to craftsmanship. Indeed, the deep social ferment of that period, instead of heralding new consumer aspirations, turned people once more to the past as a source of reassurance, and led them to identify with monumentality and often excessive volumetric bulk as a way of emphasizing social status.
The exhibition's topicality lies in this sort of pluralism of idioms, where neo-classicism coexisted with the last vestiges of futurism, novecento, rationalism and the first traces of organicism, which also prompts reflections on the neo-eclecticism that characterizes the furnishings and design of today's interiors. Some of the most significant objects in the exhibition demonstrate how young Italian architects in the thirties were determined to maintain positions that were quite detached from European design culture. This can be seen to a limited extent purely on the design level because, from the production point of view, production in large quantities was still clearly problematic.
Various excuses were used to justify the rejection of industrial production, including economic considerations and supposed justifications of a cultural-nationalistic nature. Models like Bottoni's armchair for Villa Latina (1929), equipped with a new type of suspension, were rejected by Italian industry and

produced by Thonet-Paris, corroborating the technical difficulties in which the Italian furnishing industry floundered.[2]

Young architects justified the simplified contours, smooth surfaces and abolition of fretwork and bronze encrustation with claims of practicality and hygiene. As early as 1927 Sartoris was designing strictly rationalist furniture, but the first industrial products in curved steel and pressed wood would only appear in the large projects for public buildings, such as the Gualino Building in Turin. The influence of younger architects was most visible in the organization of space in buildings most representative of the new trend. For the Casa del Fascio in Como and for the Sant'Elia Asylum, Terragni designed armchairs and chairs in tubular steel, prototypes in which linearity ruled over volume. Bottoni worked on furnishings for public housing, while Mucchi designed and produced the first "economical" furnishings. Gio Ponti, responsible for the furnishings for the Montecatini Building, formulated interesting solutions for relationships between furniture, lighting and plant design.

The 1933 Triennial demonstrated the triumph of Italian materials: securit glass, linoleum, buxus and tubular steel would become the most popular choice for rationalist furniture. When Italy enforced an economic policy of protectionism in 1936, this trend grew with studies on glass, aluminum alloys and semi-processed products, such as Masonite, hardboard and linoleum. Aluminum alloys (anticorodal, chrome aluminum and oxidized aluminum) were first used in construction projects for railings and door and window frames (such as the aluminum frames for the Montecatini Building in Milan). Their resistance to atmospheric agents, combined

with their lightness and malleability, allowed them to be combined with glass and used for ship fittings and luxury goods in general.
While aluminum alloys could be produced relatively easily in Italy by exploiting mines in Istria and the new hydroelectric plants, the glass industry, which utilized the siliceous sands of Veneto and Tuscany, was entirely domestic. This material was also the subject of many studies, with the aim of fulfilling the widest range of requirements in increasingly sophisticated ways. Refined, exquisite products were produced that had little in common with the original Empoli glass. These included termolux, produced in widths and sheets of different sizes, with various applications for thermal and acoustic isolation, filtering light, or furnishings that had to be resistant to atmospheric agents and acids. Termolux (a complex vitreous body formed of two glass surfaces that hermetically sealed a glass fiber filter) can be considered the progenitor of the current multi-layered glasses, and was used either curved or flat in furnishings with strong adhesives for fixing and protecting the more exposed edges.
The research carried out by the glass industry (Balzaretti e Modigliani di Livorno, Saint Gobain, Vis, Italor, Risana Specchi and Lastre Colate di Vetro) led to the spread of safety glass, of which Securit and Vis are the most well-known, due also to competitions held by Umanitaria and the Unione Vetrari Italiana, and by *Domus* and *Casabella*. It was used for the manufacture of furniture as well as for interior design to make partitions and dividers, developing ideas that were initially introduced for trade fair exhibition areas. The use of these materials determined a new conception of furnishings, provoking rich imaginations (fur-

niture by the Romeo firm, and Littorio glass printed with icons of the regime) and boosting the creative and experimental spirit of the new generations of architects, stimulating them to break with the traditional building shell and move towards a more composite relationship between interiors and exteriors.
Linoleum was another home-grown material, which had the advantage of being sound-absorbent, waterproof, insulating, immutable, easy to care for and hygienic. This material, manufactured by oxidizing linseed oil, cooled and mixed with pine resin and wood flour to form sheets on a jute backing, was produced in different variants: plain, marbled, streaked, inlaid, lincrusta (backed with cardboard instead of jute), and silusta, finding widespread application as a material for paneling and flooring, especially in public buildings, as well as for furniture.
Buxus was also studied and produced for this purpose. It was a sort of ossified cellulose made of rectangular sheets in different colors, which could be applied to any surface (wood, plaster, metal, cardboard) with any glue. It was shaved, scraped, filed, honed or varnished, and used in furniture making, construction, or the manufacture of leather goods, depending on the stiffness of the sheets produced.
The furnishings exhibited at these Triennials used traditional materials in unusual ways, and new steel and glass materials to produce unusual effects in stark contrast to furnishings that were essentially symbols of the decorative novecento style. The issue of furnishings dominated the popular press. Giancarlo Palanti's introduction to his *Mobili tipici moderni* collection for Edizioni Domus became the bible for furniture makers of the period.
"*The ideal home*, the type of dwelling towards which the best modern furniture

makers tend, will see the abolition of all large traditional furniture; furniture will be reduced to tables and occasional tables, chairs, armchairs, beds. All the various wardrobes, chests of drawers, buffets and sideboards will have to vanish, their material and aesthetic bulk removed and replaced by wall cupboards or small service and storage rooms, depending on the importance of the dwelling.

And it is not true that, in order to do this, it is necessary to be the owner of your home and that it cannot be done with a rented property. This concept, this objection proposed by the majority, depends on how rental property has been built up to now, and how it is still built. Indeed, the concept of the 'furniture-less home' is never as appropriate as it is when applied to people who move at regular intervals from one place to another, who will not have to drag truckloads of furniture and who will find, already arranged functionally and according to carefully considered criteria, an abundance of cupboards and storerooms for clothes, linens, dishes, etc. It is also an excellent concept for those who are setting up home for the first time (especially if they come from the less wealthy classes), who will have huge savings on the purchase of the furniture they need.

This would do away with the temporary, contrived and uninviting look that tells you from a thousand miles away that an apartment is rented. This is due to the fact that very rarely, if ever, will you find an apartment in which the furniture you already own fits perfectly; and so we have those typical makeshift solutions: sealing a door to lean a cupboard against it, sofas beneath a window, a chest of drawers obstructing the swing of a door, items of furniture lining the corridor and other aesthetic delights of this nature.

But I realize that, my task being to present a book on furniture, I am prophesying and hoping for the disappearance of furniture; this would be quite funny if the concept of the 'furniture-less house' were not destined to remain indefinitely a dream, only realizable by the lucky few who build their own houses.
Too many houses, in fact almost every single one of them, are still designed and built nowadays with the traditional idea of providing a given quantity of square meters divided arbitrarily into smaller section. Only two of these sections, the kitchen and the bathroom, have their function necessarily established a priori; and this is with the illusory intention of leaving the future inhabitants the freedom to establish the role of the various rooms as they wish. The result of this freedom is that the various rooms are not designed to be anything in particular. Every home dweller has unavoidable and heartfelt needs, such as, for example, having the kitchen close to the dining room, the bedrooms close to and easily reached from the bathroom, a larger living room than bedroom, certain rooms facing south, others facing west, others north and a thousand other requirements. The destination of the rooms should be established beforehand and designed so that the passageway between them is as logical and convenient as possible, avoiding the function of one room intruding onto another, equipping each space with the necessary cupboards and storerooms, etc.
But when (and this is just a crazy hypothesis) these healthy criteria begin to take shape in construction, we will still have to take into account that there are older buildings that will not fall down from one day to the next, and consider the attachment to old habits so deeply rooted in our bourgeoisie, who for a while will want

to drag their old furniture along with them. If necessary they will replace old furniture with new, but it will be a sorrowful wrench to part with the buffet sideboard, glass cabinet, highboy, dressing table and, who knows, faced with a modern apartment where all this is useless, perhaps, perhaps . . . they'll go in search of an older one where they can fit, even awkwardly, their habitual lumber. This is why furniture makers should not be afraid: the era of furniture is not yet dead and will not die for some time; it is and will be necessary to produce a great deal more."[3]
The various Triennial home exhibitions in this period highlight how the question of living was no longer seen merely in terms of adapting a style to the demands of "modern life," but as a problem to solve in connection with the technological potential of industry. The furniture pieces created by Albini, Camus, Palanti, Romano, Clausetti, Gardella, Mucchi, Mazzoleni and Minoletti were designed on a modular grid measuring 66 cm, held to be optimal for the various requirements of furniture. At the 4th Triennial, however, the BBPR group and other rationalist architects introduced the concepts of flexibility of use, transformability, standardization, maximum simplification and technological precision.
To summarize the evolution that furniture underwent in this decade, its transition from highly imaginative research to realistic satisfaction of precise functional needs, Albini's statement on modern taste is apt. He noted that it "has among its clear characteristics the aversion to exceptional things, to the search for novelty for the sake of it, to technical stunts, to unique pieces, preferring instead ordinary, plain materials, simple and pure technical solutions, mass-produced objects. . . ."

But even though these were the aims found in the treatises or exhibitions of the age, it is clear that custom-built furniture formed a large portion of the only truly prosperous market in the thirties – furniture for the bourgeoisie.
The famous craft workshops that sprang up in the art nouveau period from the partnership between great cabinet-makers – Quarti, Zen and Valabrega, just to mention a few – and architects, continued with their standard practice of studying ad hoc solutions for different customers, often becoming the sole producers of furniture conceived by designers who planned entire projects for interiors. On the other hand, in contexts where the furniture industry was more developed – Scandinavia, for example – a lot of production was custom during this period. Depending on the size of the order, single pieces would in due course, however, represent an opportunity for future industrial production.
As Vittorio Gregotti rightly observes, the modern in Italy was somewhat late in gaining ground. "To talk of design in terms of these furnishings is plainly improper. This is due to four factors. The notion of design as a specific activity, though methodologically connected to the modern movement, cannot be traced in Italian culture before 1940, the year of the 6th Triennial. Design oscillated in the twenties between the preoccupation of defining traditional 'applied arts' in a modern sense, and at the same time attempting a breakthrough towards spaces for experimentation that were broader and more closely linked to the visual arts. It was intent on a purely utopian idea of the world of the machine, even more than production, as formal inspiration. Above all, the most coherent frame of reference for the rationalist furniture of the thirties remained

that of architecture. This . . . was principally owing to the fact that the object was constantly destined to play a precise spatial role in furnishings . . . and was never conceived as a product: rather it was an element of an ideal or real whole in which a totally renewed global conception was micro-experimented."[4]

The furniture shown in the exhibition *Custom-Built* is everyday furniture that the upper middle classes chose for their own homes, furniture that we can only superficially reduce in a simplistic fashion to protodesign experiments. As in architecture, the modern language of furnishings was thwarted by the strong neo-classical traditionalism closely linked to the difficulty of the average consumer who, as Gabriella D'amato observes, "found it hard to think of the item of furniture as a product, the result, that is, of programmed design and mechanized production. Only in postwar Italy did the formidable requirements of new housing and the need for functional furniture to fit the restricted spaces of public housing awaken many categories of artists, manufacturers and customers. In the climate of redevelopment, designers found themselves having to respond above all to the demands of practicality and economy and, in the specific case of furniture, traditional craft workshops were forced to gradually metamorphose into small and medium-sized industries.[5]

We could now ask ourselves whether the examples provided represent a significant cross-section of the period under examination. An exhibition is always a specific choice, a critical exploration of the complexity of reality and, consequently, always one-sided. When the means at your disposal – time and funds – are scarce, it becomes perfectly clear how the relationship between historical reality and interpretative slant should be evaluated. As I do not believe in a kind of metaphysics or historical objectivity as a reassuring source of absolute certainties, I will note that the proposed slant tends to favor the Milanese experience, deliberately chosen as representative of the ongoing debate in Italy in the thirties. This meant neglecting other undoubtedly stimulating realities, but is justified by being in tune with the cultural debate reported in the press, which at the time saw Milan as an intersection for different critical experiences and the seat of a lively reappraisal of the problem of design. The exhibition deals with this problem through the specific reading of furniture in relation to the spaces for which they were conceived, without reference to a more elaborate system of elements used to define the living culture of the period.

If emphasis is indeed placed on various forms of expression oscillating between novecento, rationalism and organicism, as well as materials and manufacturing techniques, perhaps a reference to other functional domains is lacking, such as the furnishings for transport or urban spaces, or the world of objects, in particular that of decorative objects from which a unity of taste and expression would emerge. We see, in fact, furniture typologies designed for private, work and public spaces as expressions of different formal and material inquiries. For the most part, however, this furniture was conceived for the bourgeoisie and emerging entrepreneurial class that needed to portray itself in new and modern terms. For example, in the case of novecento furniture, it was obviously luxurious and high quality furniture, certainly not suitable for mass production. The return of symmetry, stylistic features like Ponti's tapering leg, and the classic forms of traditional furniture that had been re-examined and simplified, determined an adherence to classical culture that saw Milan as the center for the propagation of the new language. As Gabriella Zuco affirms, "The classical emphasizes an immobile, immortal beauty, constantly present throughout history: it is the syndrome of stability and order. It is the magnet for the new rich classes, the rising bourgeoisie and entrepreneurial class; a style that is in some way innovative but certainly not revolutionary."[7]

In some cases, such as Terragni's furniture, we see how a Gropius-style analysis can also be used to interpret the project with regard to the design/production relationship. As Raffaella Cripi rightly notes, "A certain tendency to rationalize the production process by subdividing it into the largest possible number of simple operations, each one reliant on a different worker, alienates the designer from the management of the process, jeopardizing design authorship to the point of obliterating it. But the designer belongs earlier in the process, to that area where the prototype is made and where experimental research is carried out, which is, in the most fortunate of cases, the prelude to mass production."[8]

It is there that the material dimension of the design operation was salvaged. Design returned to being the theater of concrete realities. If we understand "the act of designing" as a tendency to continuously surpass oneself and provide a means of promoting production through a renewal of forms, then many of the propositions elaborated by designers . . . be they prototypes or standard products,[9] are to be regarded as design in the fullest sense of the term, if only because they tangibly demonstrate the intention to integrate artistic quality with mass production quantity.

Notes

1. A. Pansera, *Storia e cronaca della Triennale* (Milan: Longanesi, 1978).
2. A. Dell'Acqua-Bellavitis, *Storia e cultura dell'arredo* (Milan: Cusl, 1984).
3. G. Palanti, *Mobili tipici moderni* (Milan: Domus, 1939).
4. V. Gregotti, editorial in *Rassegna*, no. 4 (October 1980).
5. G. D'Amato, *L'arte di arredare. La storia di un millenio attraverso gusti, ambienti, atmosfere* (Milan: Bruno Mondadori, 2001).
6. D. Riva and C. Tonelli, *Mobili Italiani anni Trenta* (Milan: Galleria del Milione, 1982).
7. G. Zuco, "Piero Zen tra tardo liberty e déco," in *Arredi del XX secolo* (Milan: Politecnico di Milano, Department of Architectural Planning, 1998).
8. R. Crespi, *Personaggi e storie di industrial design* (Bari: Mario Adda Editore, 2002).
9. M. Universo, *Gio Ponti designer Padova 1936-1941* (Bari: Laterza, 1989).

Between the Workshop and the Factory
"Custom-Built" Objects in Italian Design Archaeology

Silvana Annicchiarico
Curator, Permanent Collection of Italian Design
La Triennale di Milano

"Custom-built" is a conceptually ambiguous category. In a certain sense it is even *deliberately* ambiguous, commencing with the very linguistic formula that names and defines it: a "custom-built" object is, in fact, established at the outset by its professed extraneousness to a category (*mass production*) in which it does not recognize itself, but which it needs in order to define its own identity. As if to say, It is only by referring, even linguistically, to *what it is not* or *what it feels it does not belong to*, that the "custom-built" object endeavors to say something about what it is, or what it aspires to be. Its identity is therefore based on a negation, an estrangement, a distancing. It is precisely this act of naming, however, which reveals the genetic implication that the "custom-built" category dwells precisely on what it refuses to be. It is precisely by claiming to be "custom-built" that the object acknowledges that "mass production" has the regulatory and paradigmatic authority needed to define what is external or foreign to it.

To rephrase, the "custom-built" object emerges as an artifact that, though aspiring to show connotations of *uniqueness* and *exceptionality*, can only categorize itself within the borders of a territory dominated by standardization and the *hegemony of mass production.* In short, it is always *mass production* that delineates the categorical perimeter and the boundaries necessary for that act of conscious *déplacement* that is implied within the very concept of "custom-built." Which, considered in this way, proves to be something apart and completely "different" from the highly crafted product, or the unique, one-of-a-kind piece designed and made in the context of a culture that is pre-mass production (or non- or anti-mass production).

The latter is *unique* or *rare* in a scenario made up exclusively of unique or rare pieces, while the "custom-built" product strives to be so in a scenario in which uniqueness and rarity are an exception (or sometimes a foreign body, a residue, a counter-trend, a wave of nostalgia, a snobbish act) compared to the practices of production and consumption that are dominated precisely by mass production.

As we know, there is now an enormous amount of serious scientific literature dealing with these issues. Since at least the sixties it has attempted to focus conceptually on the complex relationship between what is "mass-produced" and what is "custom-built." Gillo Dorfles's observations, for example, are particularly relevant. He maintains that the detection in "custom-built" objects of the desire to "be distinguished from others through the adoption of an object – an automobile, scooter or motorboat – that is unusual, or at least owned only by few, by an elite,"[1] underlines the tendency of such object typologies to be conventional, prone to conform to the fashion of the moment, to positively pander to and emphasize it, though feigning to outdo it owing to its market *rarefaction* or lack of accessibility due to elevated cost, use of precious materials or more sophisticated finishing touches. With the intention of finding the link between craft and industrial production, Dorfles also notes that it is impossible to pinpoint what is "custom-built" and what is produced in a "limited series." For the Triestine scholar the concept of a "limited series" evokes, in fact, two diametrically opposed object typologies: on one hand, those objects that – though produced by a method that falls squarely into the category of industrial design – are "destined for an extremely limited mass production,"[2] such as large jet-propelled aircraft, or ships, submarines, locomotives, cruise ships and so on; on the other hand, those objects that – according to a habit that Dorfles judges to be "socially reprehensible"[3] – exploit industrial manufacturing's ability to replicate the object, but are limited as to demand and production in order to endow the object with a connotation of uniqueness or sheen of excellence. This category includes high fashion products, but also furnishings, knick-knacks or "designer furniture," where slim demand and deliberate scarcity of supply go hand in hand with "the high price and particular affectation of the merchandise."[4] According to Dorfles's analysis, the limited *mass production* of objects of the first type is owed to an inevitably small and "super-individual" customer base, while for objects belonging to the second type the reason is a desire for distinction that rejects mass production though trying to gain some advantage from mass production methods. In both cases we are nevertheless far from the idea and practice of craftsmanship, moving instead in an ambiguous – but for that reason extremely interesting – border territory in which the acceptance of mass production cohabits with its negation, or with the tendency to transcend the generative matrix from which certain objects have been created. Paradoxically, precisely because of its unethical and problematic make-up, the notion of "custom-built" proves to be extremely useful and profitable in order to trace what we might define as the *epiphany of design* in modern Italy's social and industrial history. This, at least, is the original idea that triggered the research that led to this exhibition.

Certainly the notion of "custom-built" seemingly *grates* and *rubs against* the need for standardization and uniformity that lies at the basis of design. If it is true that design "considers the object as a product, independently of its placement and context,"[5] a "custom-built" object instead refuses to allow itself to be used in a ubiquitous and undifferentiated fashion and – above all in the case of furniture – is made to a specific order and dependent on a set architectonic, spatial and social context. It is made for *that* house, for *that* environment, in tune with *that* taste. It cannot be replicated, transferred, standardized. And yet it has something to do with industrial production. It is tied up with the theory and practice of design at the very same moment it is differentiated from it. It is precisely in its being different that it reveals something of itself and the world. It speaks. It transmits symptoms and signals.

The objects brought together and presented in this exhibition all belong to this category: they are unique pieces, prototypes or models of a "limited series" made on commission for a specific need or requirement. From a strictly philological point of view, they do not perhaps wholly belong by right to the history of Italian design. And yet they have something to do with this history. Intimately, profoundly, productively. Either because they indicate design and creative solutions that production will appropriate when industry determinedly decides in favor of mass production, or because they manifest – especially those manufactured in an already mass-produced age – a sort of nostalgia for the unique piece as a form of romantic resistance to the standardization of industrial production. Some of these objects are precursors of the future; others fantasize about and grieve for the past. Some anticipate the times; others – to borrow Walter Benjamin's famous metaphor – express a barely concealed *nostalgia for the aura* of the age of technical reproduction of objects and artifacts.[6] And yet, even when they express a tension towards uniqueness and inimitability, these objects provide a precious testimony of the way in which the culture of industrial design has imposed and affirmed itself in our country: a sequence of thrusts and counter-thrusts, spurts and inertia, forward surges and rapid retreats. They help to trace a map characterized by imprecise and contradictory zigzag contours, never ascribable to unambiguous schemes or easy formulas solved once and for all. The intention of this exhibition is to investigate this contradictory *modern sign system* using a symptomatological approach that makes further contribution to the project of shedding light on the archaeology of Italian design, in the belief that no history – including that of industrial design – can ever be represented in a linear fashion, but involves the continuous exploration of layers and sediments, deposits and finds that incessantly call into question established patterns and acquired certitudes.

In the Italian social-cultural context, it is well known that one cannot properly speak of design before 1940, when the International Exhibition of mass-produced goods, curated by Giuseppe Pagano for the 7th Triennial, placed the question of mass production at the center of attention. Previously, from the post-World War I period and for the whole of the thirties, we witnessed "the attempt to define in a modern sense traditional craft production, to liberate it from "revival" trends and to attempt to bring it more in line with progress made by the visual arts, from the avant-garde onwards."[7] If it is true that in the mid-thirties, the nostalgic and pro-aristocratic passions of Ugo Ojetti (fervently intent on defending "decorative arts" from the humiliation of the "democratic chimera" through an appeal for "necessary luxury" in the production of furniture and objects)[8] found a ready response in the modernist and democratic positions of Edoardo Persico and in his flat demand for "pride in modesty" as an inspirational criterion of production,[9] it is just as true that for the entire decade frequent allusions to the need for standardization remained for the most part dead letter in the face of an industry not yet able to handle the requirements of the modernization now spreading throughout society. In fact, objects produced in this period are at most craft prototypes of future mass-produced products, but do not by any means foreshadow a model of production of quality objects for the widest range of consumers and for mass consumption. From this point of view, the thirties reflect – though we cannot ignore the élan of some personalities who were particularly sensitive to the demands of rationalization, modernization and democratization of production – the delays and contradictions of a country that did not yet appear mature enough to assemble and appropriate the demands of "consumer democracy" that potentially accompany modernity. As Vittorio Gregotti noted, the rationalist object in Italy in the thirties and forties "was constantly destined to have a precise spatial role in furnishings . . . and hardly ever proved to be conceived in terms of product."[10] If design was characterized, especially from the post-war period, by its ability to design and make a-topical and a-geographical objects, freed from their subjection to a place or a customer base, and able to offer itself to a plural and multivalent use, Italian industrial design archaeology seems characterized instead by objects and furniture that were created because targeted and destined for a precise place and a specific use that could not be imitated or replicated, at least on paper. For example, consider Giuseppe Terragni's furnishings for the Casa del Fascio in Como (1935-36), Luigi Caccia Dominioni's furniture for a house in the Bergamo district (1941), Luigi Vietti's furniture for the Casino di Caccia in Cressa (1941), or Giuseppe Pagano's furnishings for Bocconi University (1942). The bourgeoisie was the principle target of this type of production, as the examples of homes presented respectively at the Triennial in 1930 (the "Domus Nova" vacation house by Ponti and Lancia, the "Casa Elettrica" by Figini, Pollini, Frette, Libera and Bottoni and the "Casa del Dopolavorista" by Lovarini) and at the Triennial in 1933 (the "Casa del sabato per gli sposi" by BBPR and Portaluppi, the "Villa dell'aviatore" by Scoccimarro, Piero Zanini and Midena, the "Villa-studio per artista" by Figini and Pollini, the "Steel Structure House" by Pagano, Albini, Camus, Mazzoleni, Minoletti and Palanti, the "Casa di un artista sul lago" by Terragni, Lingeri and Cereghini) show so eloquently, all of them inspired by a "studied relationship between architecture and furnishings, fixed and moving elements, internal and external space."[11]

In fact, as Renato De Fusco has persuasively demonstrated, "The models of Pagano, Terragni and other rationalists never went beyond prototypes and never became mass-produced objects; their experience was limited to some furnishings and above all outfitting of exhibitions or commemorative events."[12] Suspended between modernist temptations and anti-modern resurgences, Italian *proto-design*[13] appears to be a composite and contradictory galaxy. On one hand, it refined pioneering and extremely advanced solutions (for example in the fields of aeronautical and automobile design, or in the sector – crucial for the regime – of mass communications, from the radio to the telephone, with the fashion of "white telephones" launched by Mario Camerini's film). On the other hand – especially in sectors with high levels of social interaction like furnishings – it ran into delays and resistance, which only the painful and difficult postwar period would manage to dispel and leave behind. At that time, a series of articles by Magnaghi published in *Domus,* and especially the expe-

rience of RIMA (Riunione Italiana Mostra Arredamento) with the exhibition *Arredamento popolare* organized for the Triennial of 1946,[14] began to draft in Italy – to reiterate Enzo Mari's famous and stimulating formula – "the potentially positive model of the standard, that is identical objects for a society of equals."[15]
The *Custom-Built* exhibition intends to contribute to the mapping of this diversified and complex territory. Without claiming to be exhaustive, but using as complicated an approach as possible, the aim is to investigate a historical period that the present Permanent Collection of Italian Design of La Triennale di Milano does not cover (the collection notoriously covers the period between the immediate aftermath of the Second World War and the end of the twentieth century), with the intention of taking a census of the objects and artifacts produced in the incubation period of Italian design, which can still be found in company archives and above all in private collections and homes. If the exhibition *La Città e il design* organized by La Triennale in 2001 mapped the company and museum deposits that legitimized the idea of a reticular and polycentric design system, vastly spread over the Milanese metropolitan territory, the exhibition *Custom-Built* – though on a much smaller and less ambitious scale – is animated by the will to carry out an initial exploration of *proto-design* finds, and to galvanize the sensitivity and readiness of those who conserve them, in view of the imminent and desirable establishment of a Museum of Design.
The objects chosen – collected by the curators and placed at the public's disposition thanks to the generous collaboration of private collectors and enthusiastic admirers – all date to the period between the beginning of the thirties

and the end of the forties. They are mostly items of furniture designed ad hoc by designers such as Bega, Portaluppi and Gardella, but also prototypes not destined to go into production (Franco Albini's floor lamp), or objects made at the request of a specific customer and only put into production much later (the *Moka* chair designed in 1939 by Asnago and Vender for the Bar Moka of Milan, which was only put into production in 1985 by Flexform). They are therefore all "custom-built" objects. But, as we said at the beginning, they were all deeply involved in the complicated process that led to the creation of the culture and the practice of mass production in our country.
All the exhibits are "original." Nothing has been rebuilt ad hoc. To look at them again today, to a greater or lesser extent scored by age and use, they simultaneously transmit an idea of absolute contemporaneity – due to the precision of their composition and design and the efficiency with which they solve the relationship between form and function – but also their inevitable obsolescence. Each one of them has something that concerns us, but also something that no longer belongs to us. There is the dream of a form of comfort that worked with volumes and bent materials (perhaps also those imposed coercively by the autarchic regime)[16] in search of perfect ergonomics, but still did not expose itself to the risk and vertigo of baring itself to an authentically social and collective use stripped of hierarchy. There is the rejection of the predominance of decoration united with the unexpected comeback of the habit or vice of ornament. There is the unresolved oscillation between the *workshop* where they were created, and the *factory* towards which they were headed. There is – especially in

certain cases – a residue of the monumentality of the bourgeois salon, which went hand in hand and clashed with the dream of lightness we had already glimpsed and sensed to be necessary, but had not yet attained, not yet totally achieved. In this innate contradiction, all these exhibited pieces end up being something more than simple furniture. They are "aphorisms" and metaphors: signs and episodes exemplary of the tortuous way with which not only industry but the whole of Italian society has tried to elaborate its labored but ineluctable relationship with modernity.

Notes

1. Gillo Dorfles, *Introduzione al disegno industriale. Linguaggio e storia della produzione di serie.* (Turin: Einaudi, 1972), p. 60.
2. Ibid., p. 62.
3. Ibid.
4. Ibid.
5. Roberto Rizzi, "L'arredamento razionalista fra artigianato e industria, in *Artigianato e Razionalismo italiano,* exhibition catalogue (Monza: MIA, 1997), p. 5.
6. Walter Benjamin, *L'opera d'arte nell'epoca della sua riproducibilità tecnica* (Turin: Einaudi, 1966 [1936]).
7. Roberto Rizzi, op. cit., p. 5.
8. Ugo Ojetti, "Lettera a Giovanni Ponti sul lusso necessario," in *Pegaso* 1 (August 1933), pp. 97-99.
9. Edoardo Persico, "Il gusto italiano," in *L'Italia Letteraria* (28 May 1933).
10. Vittorio Gregotti, "Introduzione a Il disegno del mobile razionale in Italia 1928/1948," in *Rassegna* 4 (October 1980), p. 5.
11. Maria Cristina Tonelli, "Il razionale nel mobile italiano," in *Rassegna*, op. cit., p. 10.
12. Renato De Fusco, *Storia del design* (Roma-Bari: Laterza, 1985), p. 260.
13. The neologism was coined by Anty Pansera on occasion of the exhibition *Gli anni Trenta: arte e cultura in Italia*, organized in Milan in 1983. See the essay in the catalogue *Il protodesign italiano: le origini del disegno industriale* (Milan: Mazzotta, 1983).
14. The aftermath of the First World War reveals an attitude which remains ambivalent and vacillating regarding the very concept of mass production. "Research, mainly addressing the field of sectional furniture and mass production, is not tackled by everyone with sufficient incisiveness, leaving in many cases the matter of the low cost of the furniture, which perhaps constituted the principle aim, unsettled." (G. Bosoni, "La ricostruzione edilizia e l'arredo popolare," in V. Gregotti, *Il disegno del prodotto industriale. Italia 1860-1980* (Milan: Electa, 1982), p. 290).

15. François Burkhardt with Juli Capella and Francesca Picchi, *Perché un libro su Enzo Mari* (Milan: Federico Motta Editore, 1997), p. 117.
16. On the paradoxically positive function of the autarchy in getting Italian industry to experiment with new materials, cf. Andrea Branzi, *Introduzione al design italiano. Una modernità incompleta,* (Milan: Baldini & Castoldi, 1999), pp. 98-102.

Franco
Albini

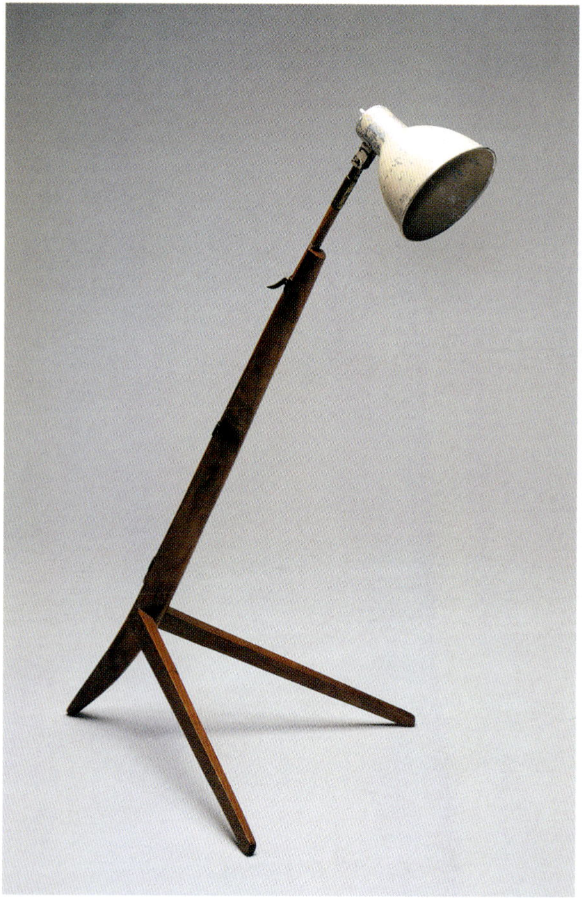

1938-1940

Mitragliera lamp in different prototypes from left:
Poggi Collection, Pavia
Fried Collection
Albini Collection

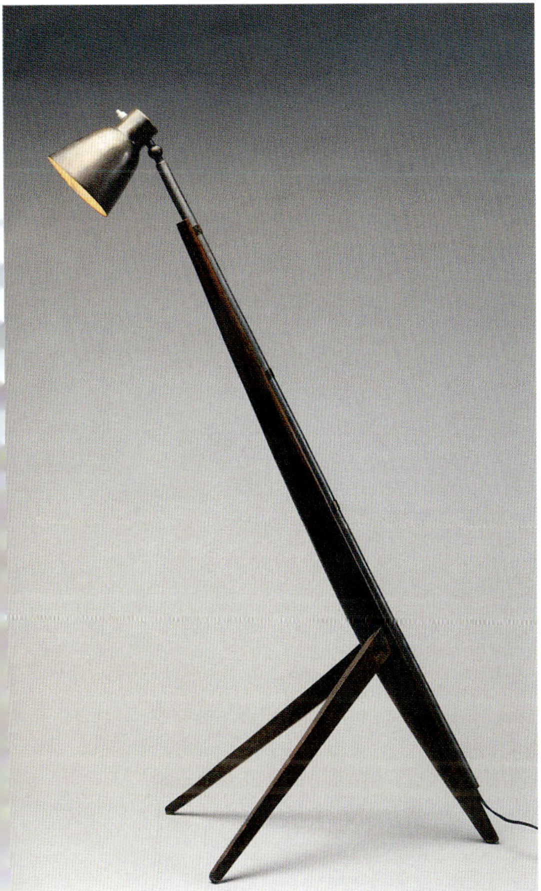

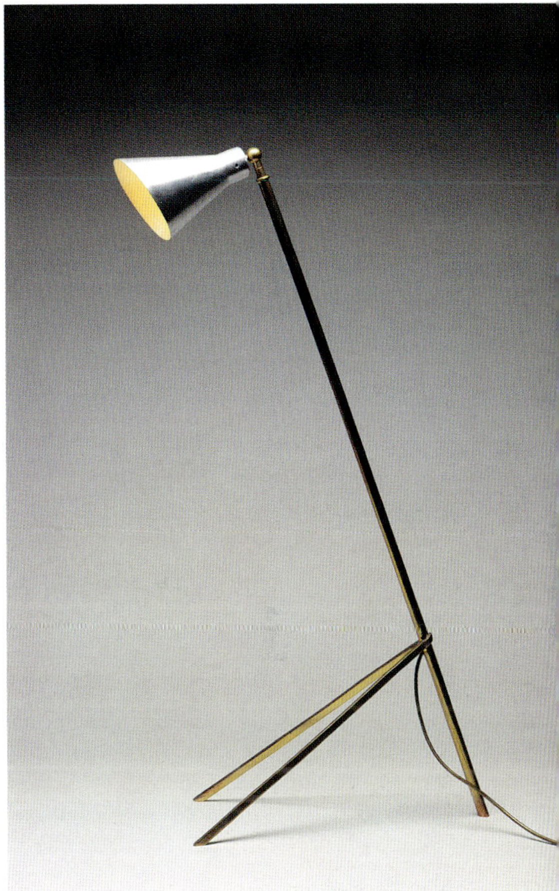

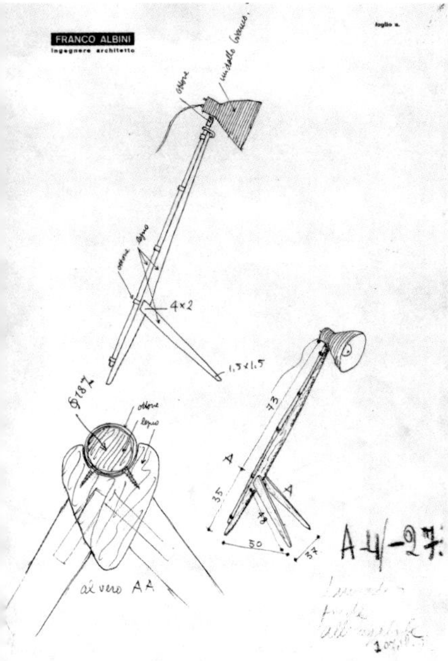

Original drawing by Franco Albini on letterhead, n.d., pen on paper, 21.0 x 29.7 cm.

The three different models of the *Mitragliera* lamp shown here are an extraordinary example of how Franco Albini was systematically working on the notion of *type* as early as the end of the thirties, modifying materials or details each time to produce different prototypes that are variations on the same design project in an incessant search for new creative possibilities. This was truly the dawning of the culture of mass production, perceived methodologically as an intrinsic requirement of design, though not yet adopted in practice by production.

The first model, probably the original creation, is made entirely of wood. The bent and hollow stand, which recalls the shape of a gun, contains and visibly supports a sliding rod circular in cross-section, also made of wood, with a longitudinal groove that enables a tongue to block the rod itself at the desired height. The reflector is connected to the rod by a square metal plate. Honed and tapered like the barrel of a gun, angled as though taking aim, the lamp switch is on the hood, which stands on a wooden structure that recalls the classic tripod of a machine gun.

The second model has the same hollow wooden stand, but a more polished finish, and its inner rod is in iron, no longer sliding but fixed. The reflector is connected to the rod by means of a spherical joint that allows it to be directed.

Finally, the third model is entirely made of brass with

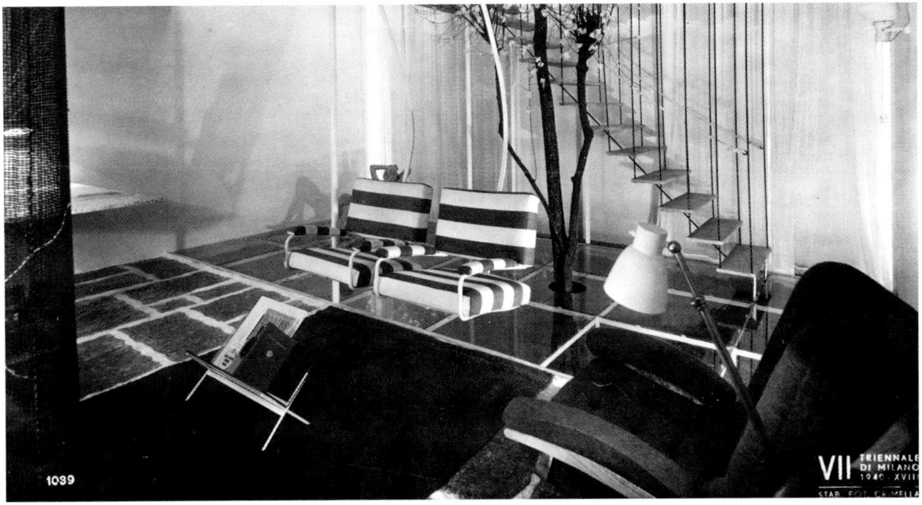

an inclined circular tube that stands on two square feet. The smaller satiny reflector is connected to the rod by means of a spherical pin that allows it to be rotated and blocked by means of a simple screw. Two of these models appeared in the living room of Albini's apartment in 1938, while one appeared in "Living Room for a Villa" at the 7th Triennial in Milan in 1940.

The lamp shown in the context of "Living Room for a Villa" at the 7th Triennial in Milan, 1940.
© Triennale di Milano Archives.

Mario Claudio
Asnago and Vender

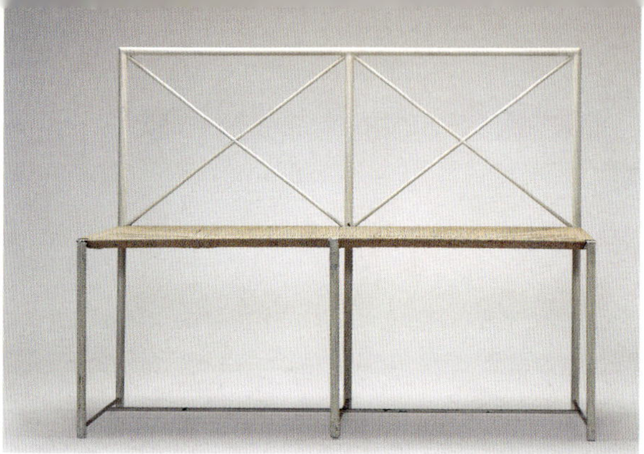

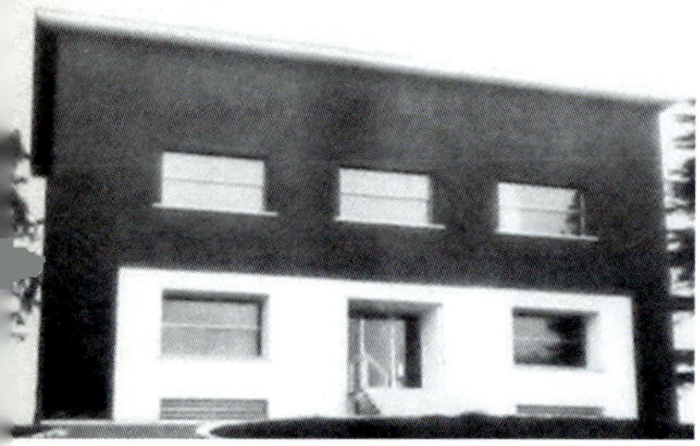

1934
small sofa and table
for Villa Marelli, Cantù
Private Collection

Made on commission to furnish a genteel residence in Brianza, these two pieces express the taste for abstraction and geometry that characterize the work of Asnago and Vender. In their search for a slender aesthetic, they stripped objects down to show off their structure and reduce them to their skeletons.
In the small two-seater sofa, as in the two-leafed table – presented at the 6th Triennial of Milan in 1936 – the metallic structure is reduced to a minimum thickness and combined very naturally with other materials (glass and upholstery fabric), according to a formal archetype that the two architects would propose again in numerous forms for various subsequent designs. Though not yet a true series, the consistent application of a cross shape inserted into a rectangle delineated a design model based on the "variations on a theme," which already evoked the idea and practice of mass production.

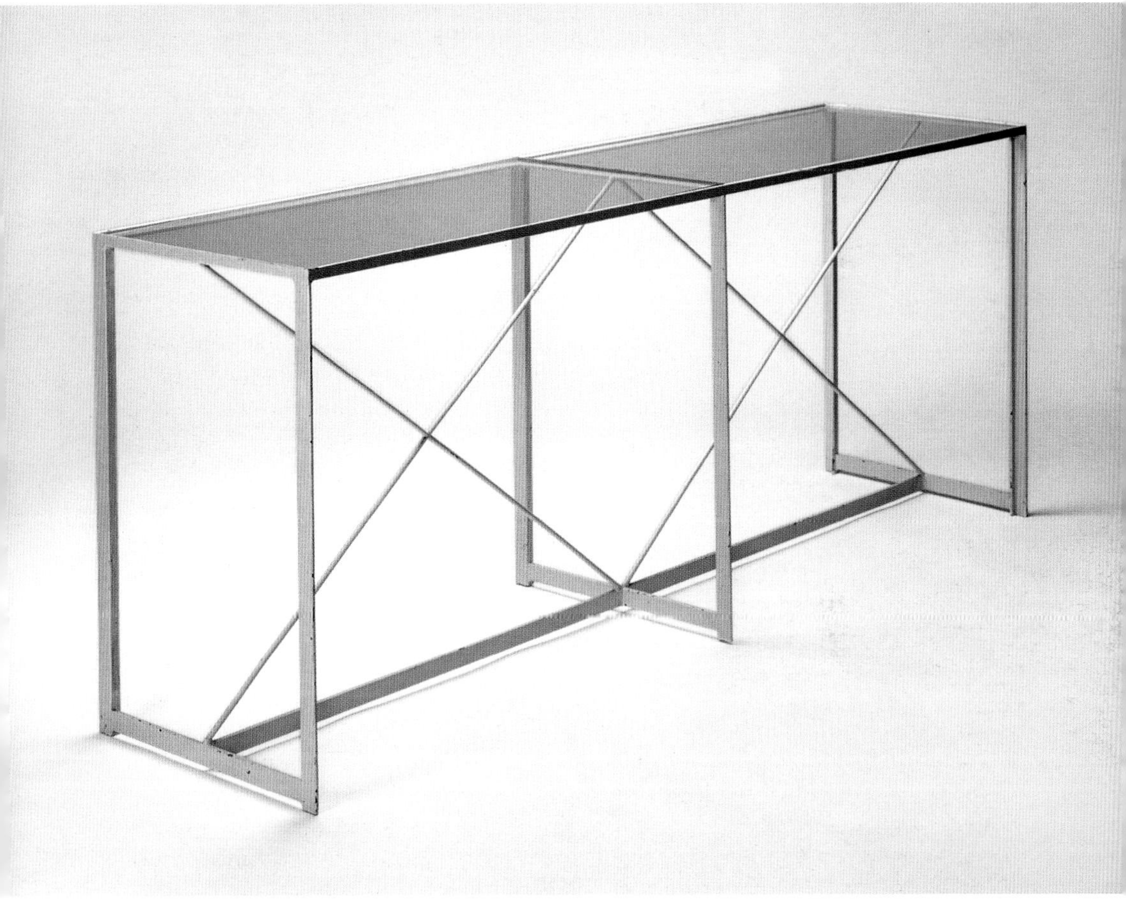

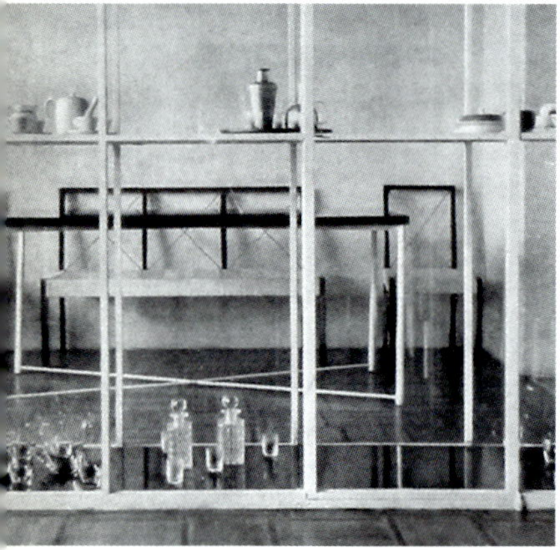

Clockwise:
The evolution of the chair in Casa Simonotti, Pasticceria Panarello and Bar Moka.

1939

chair for Bar Moka on via Tommaso Grossi, Milan
and for Pasticceria Panarello on corso Magenta, Milan
Alessandro Pedretti Collection

"Their furniture appears to be made of thin layers joined by tense filaments." This judgment from Raffaello Giolli beautifully sums up the formal work of Asnago and Vender, which finds an almost perfect expression in this chair conceived in the late thirties for public locations in Milan. The rectangular or diagonal metal intersections create a scaffolding effect that is simultaneously fragile and solid, in which the precise slotting of the small iron strips, tubes and rods represent a desire to use materials to give the objects structure, but at the same time to break them up and reduce their bulk to a minimum, producing a surprisingly light effect.
The modernity of this prototype is such that industry appropriated it, transforming it to all effects and purposes into a standard product. In fact, the *Moka* chair has been produced by Flexform since 1985.
"Already in their most famous furniture . . . originally designed for the Simonotti House in 1935 and then proposed again in numerous variants in subsequent furnishings (at the 6th Triennial in 1936, in the Moka Bars and Panarello Pastry Shop in 1939), a graphological reading reveals the same linguistic ellipsis in their architecture. The search for an extreme fragility and aristocratic simplicity of materials cancels out in these furnishings that taste for volume, cornices and inlays typical of twentieth-century style . . ."
Cino Zucchi, *Asnago e Vender. Architetture e progetti 1925-1970* (Milan: Skira, 1998), p. 189.

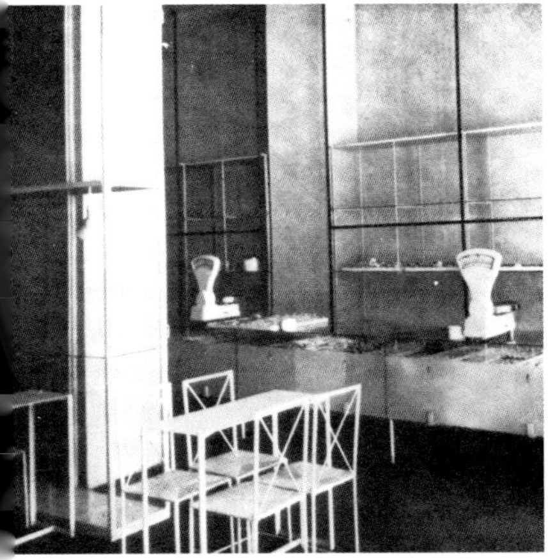

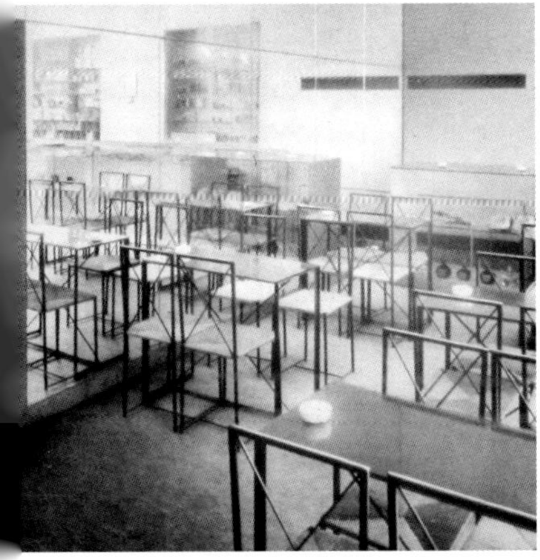

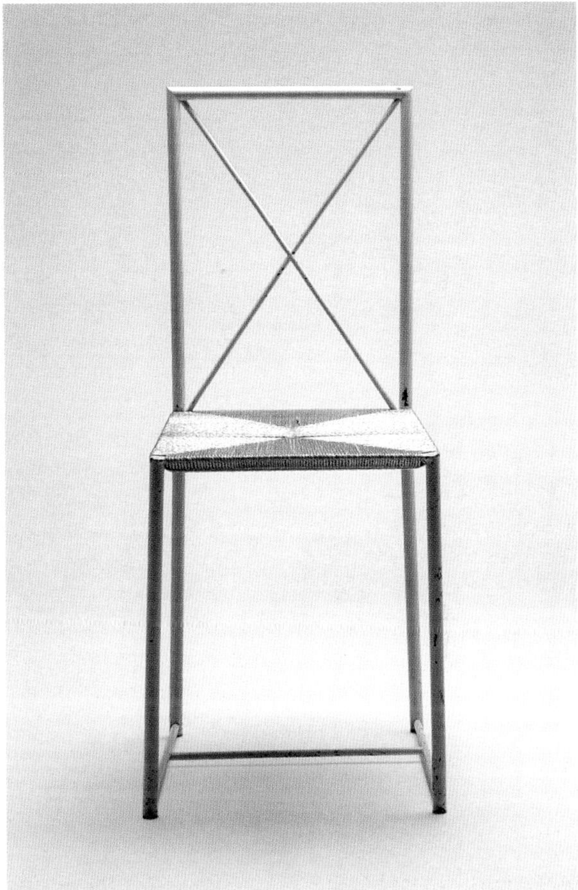

Melchiorre
Bega

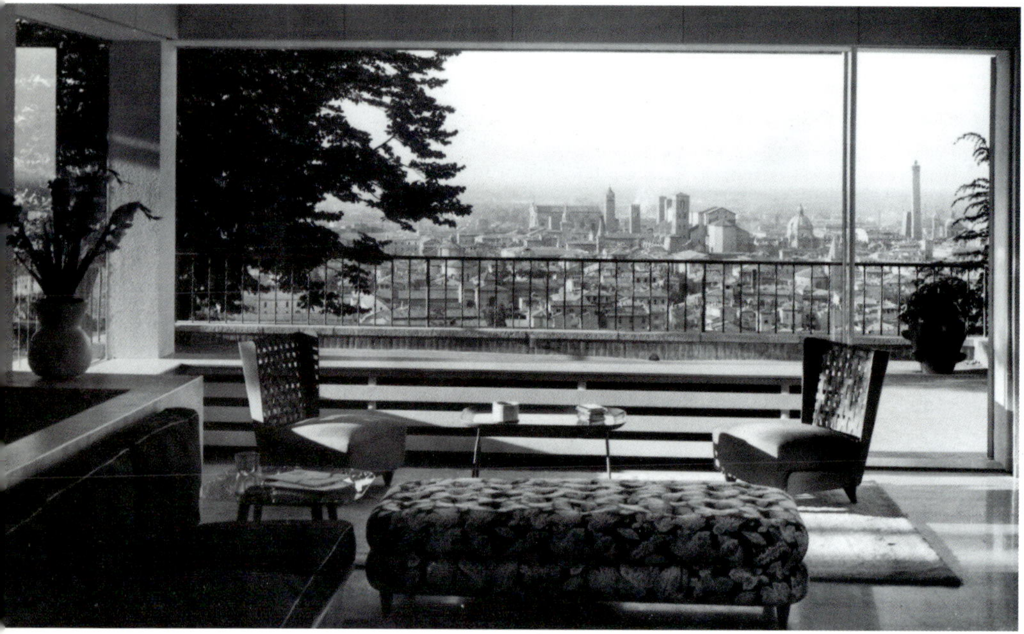

1940
armchairs and console
for Casa Bega
at Osservanza, Bologna

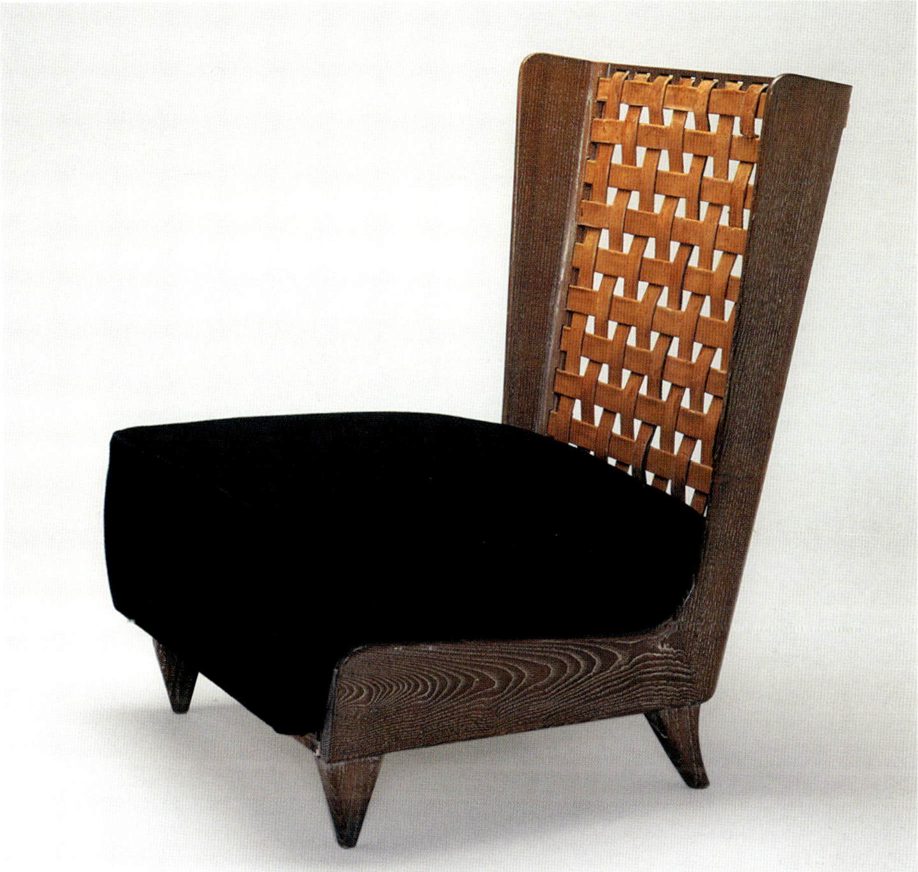

The Bega furniture factory was founded by Melchiorre's father, Vittorio, at the beginning of the last century. A new factory under the management of his children and employing 250 workers was inaugurated in 1923. Bega quickly became the trusted furniture maker of famous architects and large industries like Motta, Perugina, Spagnoli, Tradate and Unica. The company remained operative until 1976. Melchiorre Bega devoted himself to architecture and furnishings and came to Domus *in 1941 as an exponent of rationalist orthodoxy.*

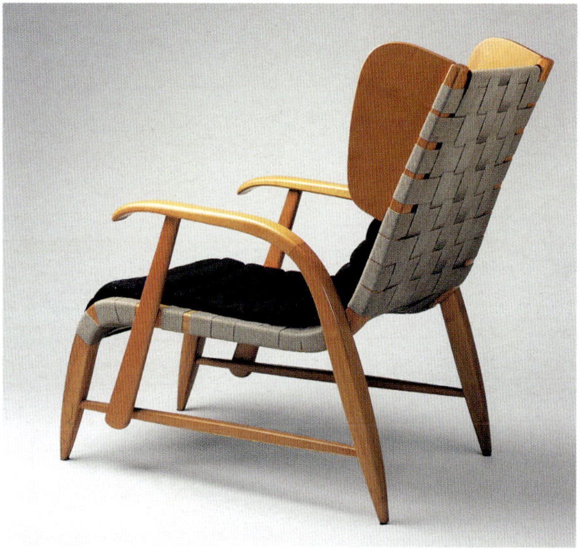

At left, the chair shown in the context of Bega's office, in Domus, *no. 165 (September 1941).*

Vittorio Bega Collection

In these prototypes, made especially for his home in Osservanza, Bologna, Melchiore Bega boldly used wood, straw, bamboo, rope, fabric and braided leather, producing vastly different results. The armchair – set in the living room with large windows overlooking the city of Bologna – is extremely elegant owing to the sophisticated work on the reticular weave of the backrest in curved plywood, and the bold modernity of the seat height, while the other two prototypes have an almost ethnic spirit.

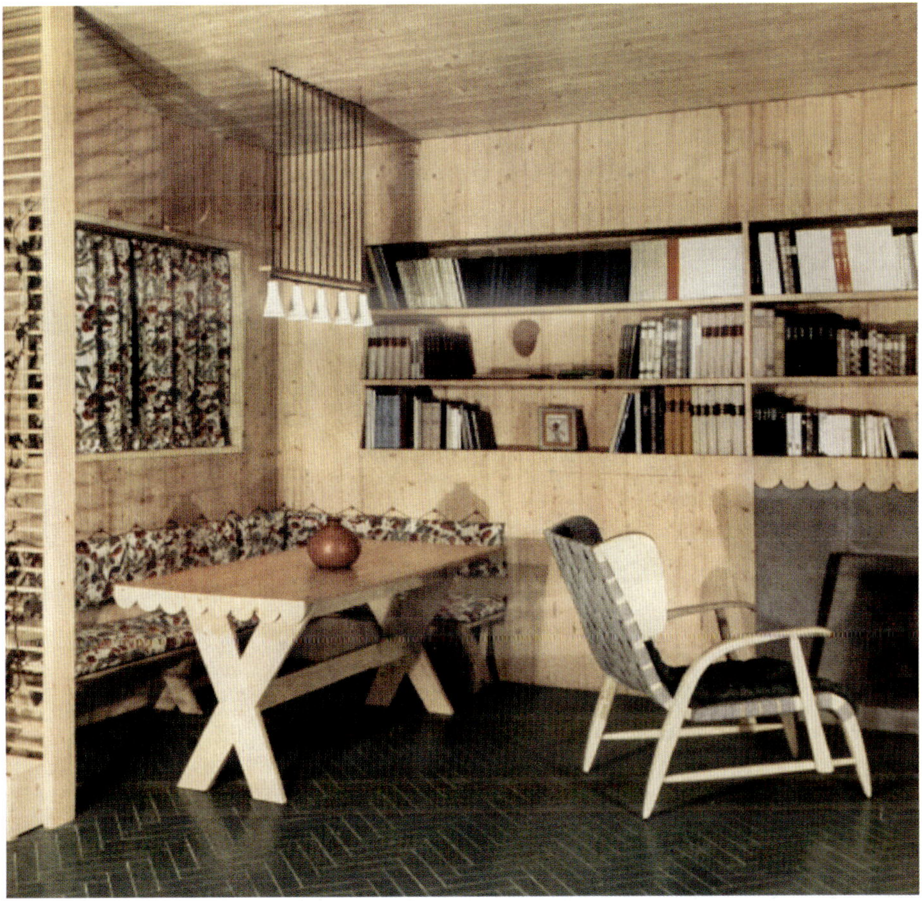

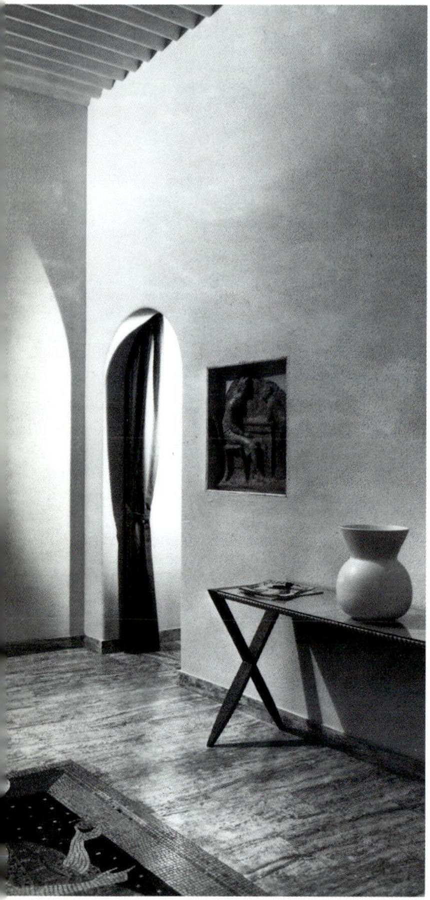

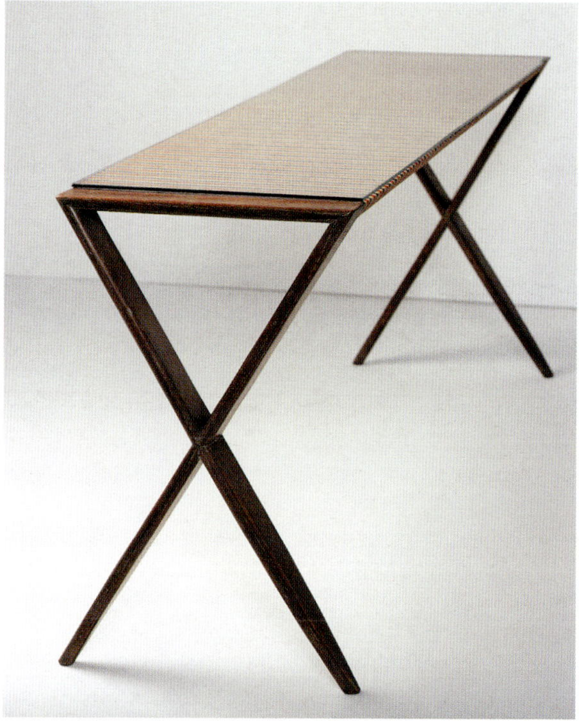

The console, with rectangular, flat legs, is made of sandblasted oak, with a very narrow rectangular top, upholstered in red and ivory striped cloth (the same colors used for the drapes), and overlaid with a sheet of glass.

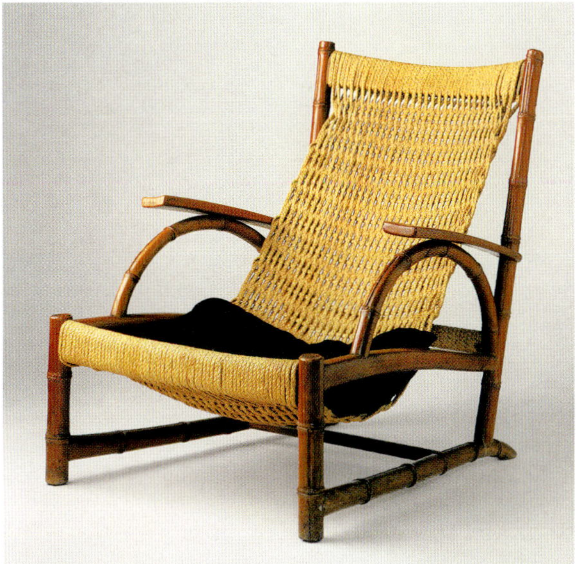

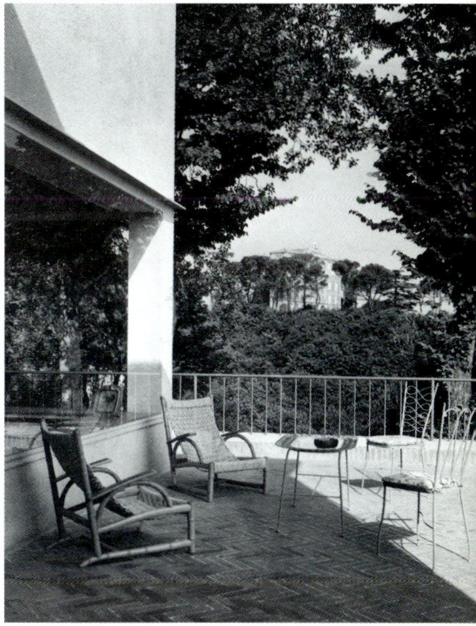

“Always naturally elegant, he considered style to be a hereditary quality of the modest class of bourgeois artisans from which he descended, and which has an ancient tradition in Bologna. It seemed logical to him that craft should evolve into industry, provided that mass production did not neglect quality materials, rational design and meticulous workmanship.”
Giulio Carlo Argan, introduction to Stefano Zironi, *Melchiorre Bega. Architetto* (Milan: Editoriale Domus, 1983), p. 6.

Piero
Bottoni

Piero Bottoni, Dressing table,
Casa Minerbi, Milan, 1930.
© Piero Bottoni Archives,
DPA, Politecnico di Milano.

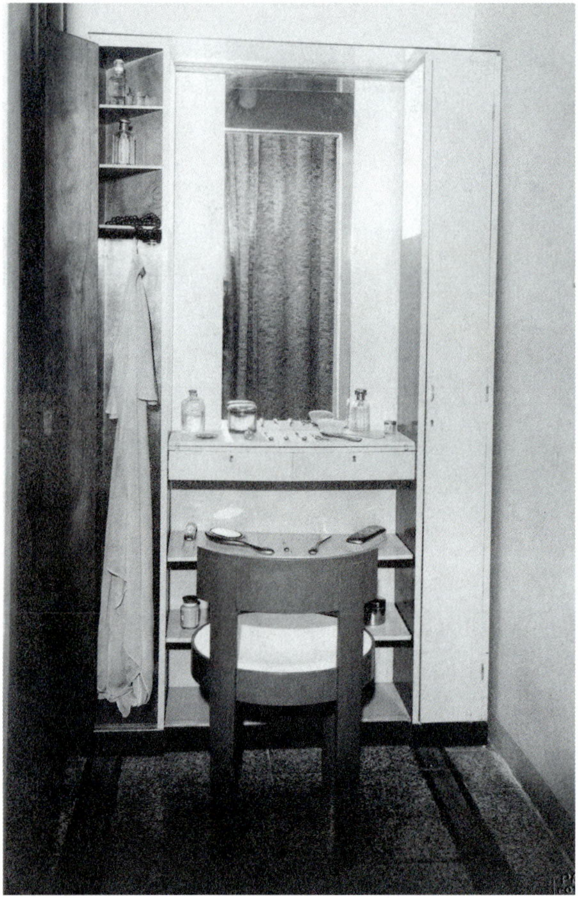

1930
chairs and table
for Casa Minerbi, Milan
Private Collection

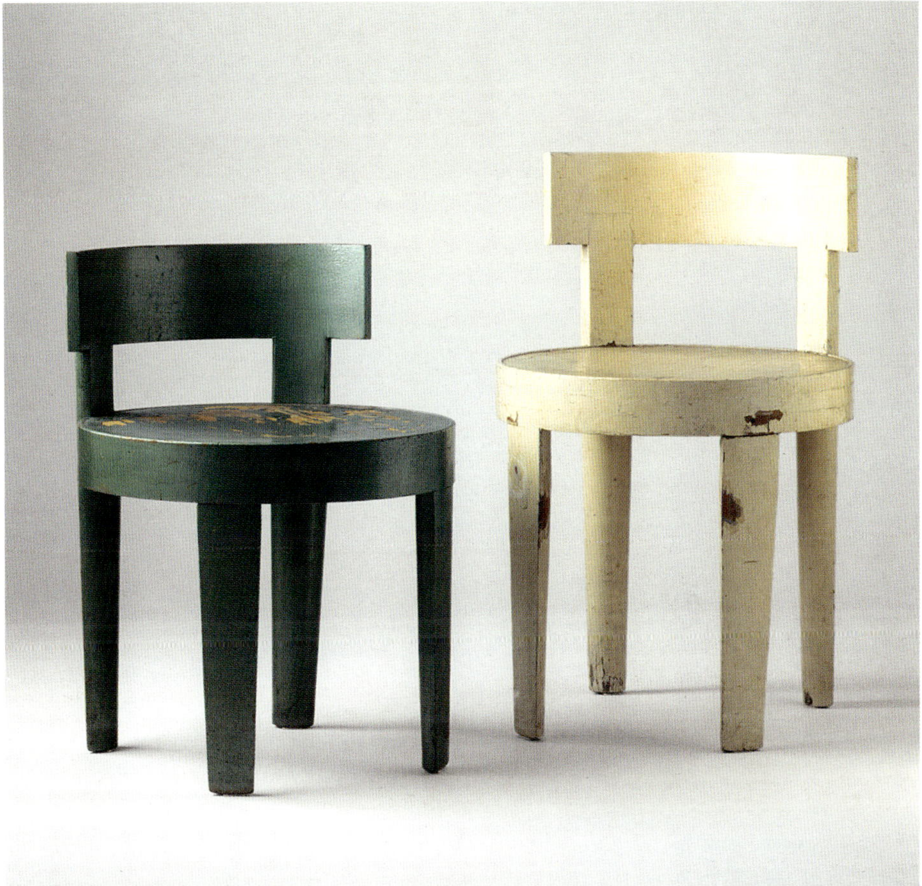

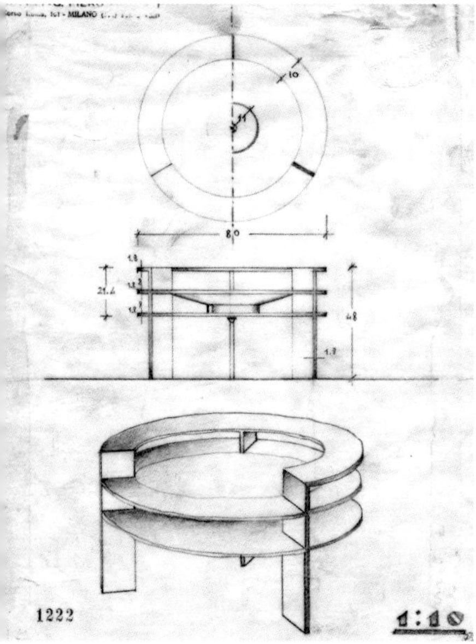

Piero Bottoni, luminous table, drawing in 1:10 scale, n.d. Pencil on tracing paper, 28.0 x 21.0 cm.
© Piero Bottoni Archives, DPA, Politecnico di Milano.

At the end of the twenties, one of Piero Bottoni's first clients was his friend, the lawyer Pietro Dello Strologo, who commissioned furniture for the dining room, music room and smoking room of his bachelor's apartment on via Montenapoleone in Milan. The table and chairs designed on this occasion (in which Giancarlo Consonni finds "a vague clownish clumsiness caused by the wide wooden fascia") were used again in 1930 for the dining room of the Minerba House, whose furnishings earned Bottoni a flattering reception from Edoardo Persico and other critics. Consonni recalls: "Persico discerns in Bottoni's furniture an ability to hold relationships with space that is typical of sculpture, while Tibaldi Chiesa perceives how the coherence of the composition extends to each piece of furniture in the room, in which colors and lights, no less than contours, lend substance to a central idea." (Giancarlo Consonni, "Arredamento di Casa Minerbi a Milano, 1930," in *Piero Bottoni: opera completa* (Milan: Fabbri Editore, 1990). In fact, Bottoni worked on *types* and stated them in a very modern way. With *chair types* he played with measurements and proportions: the backrest could be raised in classic fashion or lowered to allow users to rest their lower backs, as though it were a comfortable stool. The legs were designed with a compass to be a perfect semicircle, or else they tapered into a trapezoid shape. The wooden rails of the seat were made using

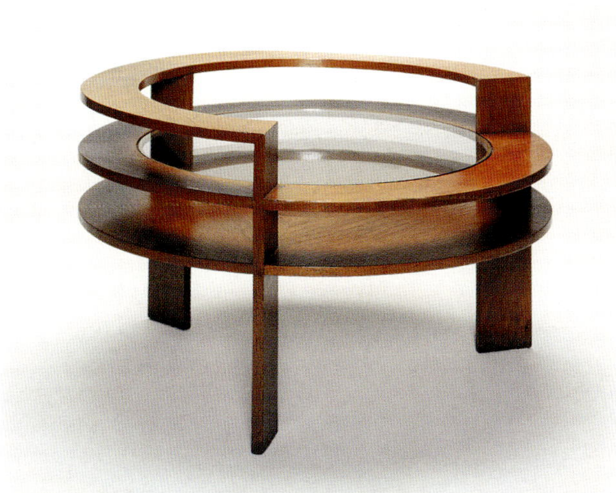

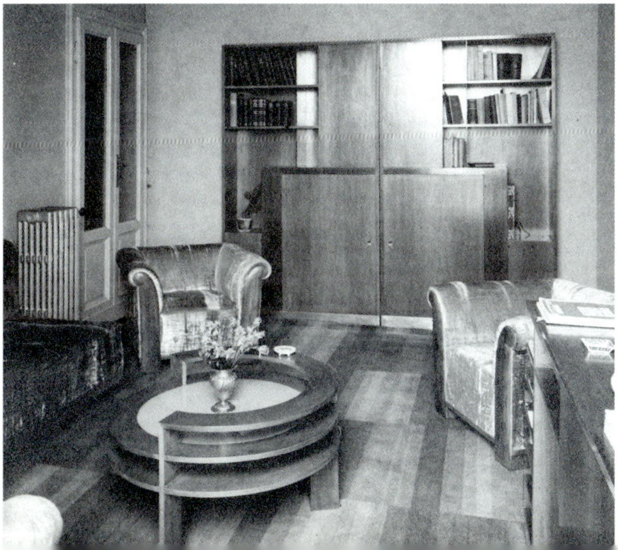

Living room, Casa Minerbi.
© Piero Bottoni Archives,
DPA, Politecnico di Milano.

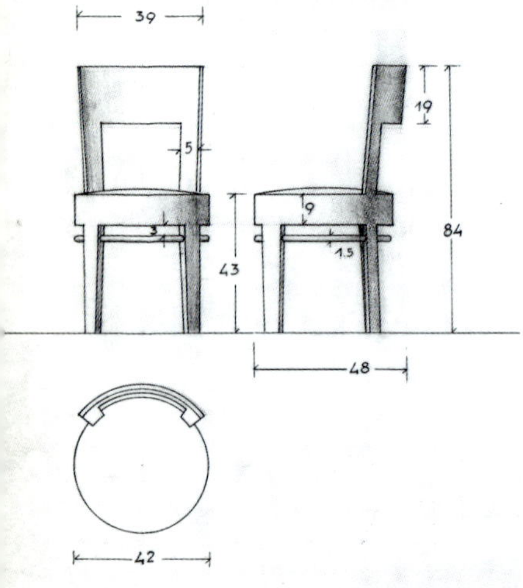

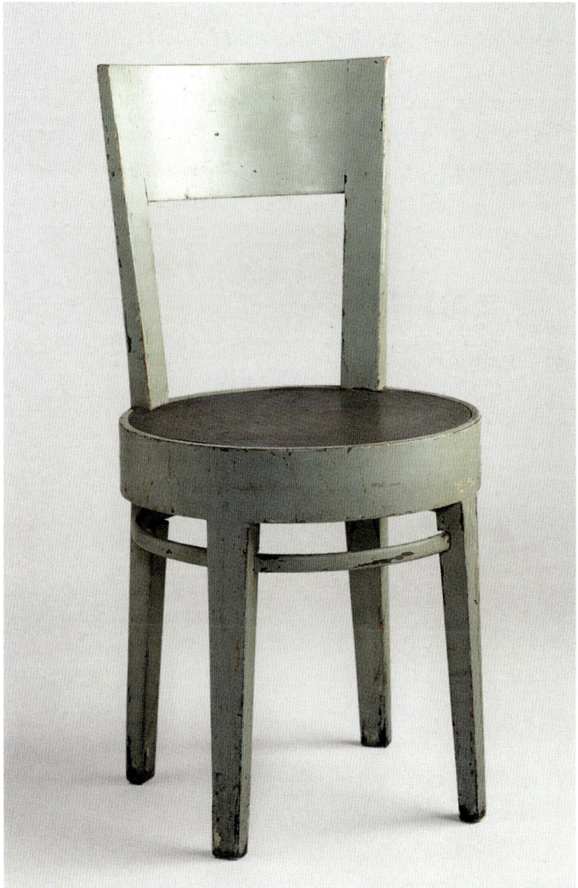

Project for country house dining room chairs, one of the "Four Vacation Houses" designed with E. Faludi and E. A. Griffini, 5th Triennial of Milan, 1932-33. Indian ink and pencil on tracing paper.
The chair was not produced for the exhibition, but on private commission.

different measurements each time, sometimes bulkier, sometimes flimsier, dividing elegantly into a fascia and a ring, separated by a deliberately empty space. In the model presented the same year for "Casa Elettrica," presented at the 4th Triennial in Monza, Bottoni again went to work on types, proposing (with Figini, Pollini and Libera) a stool and kitchen chair whose seat was so deep it could be used as a container for shoe brushes and polish.

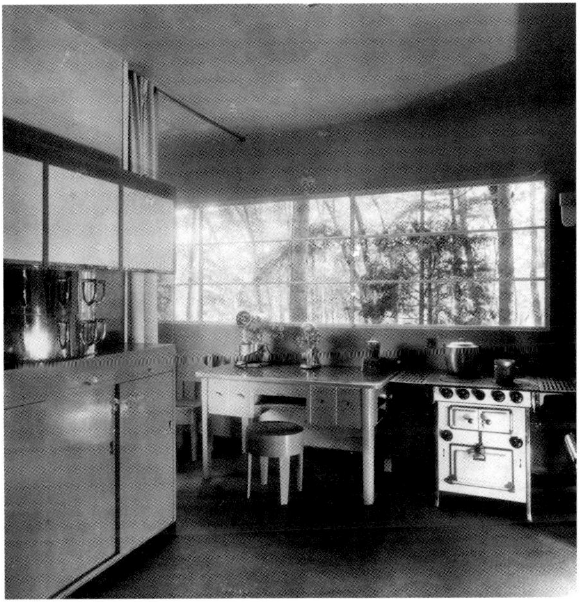

The kitchen of "Casa elettrica" created for the 4th Triennial, 1930.
© Triennale di Milano Archives.

Luigi **Caccia Dominioni**

This lamp used by Caccia Dominioni for the living room of the Bergamo house in 1941 had already appeared – in a version that does not have holes in the upper hood – in furnishings prepared by Gian Luigi Banfi for his Milanese house in 1939 (See Rassegna *2, no. 4 (October 1980), pp. 84-85.)*

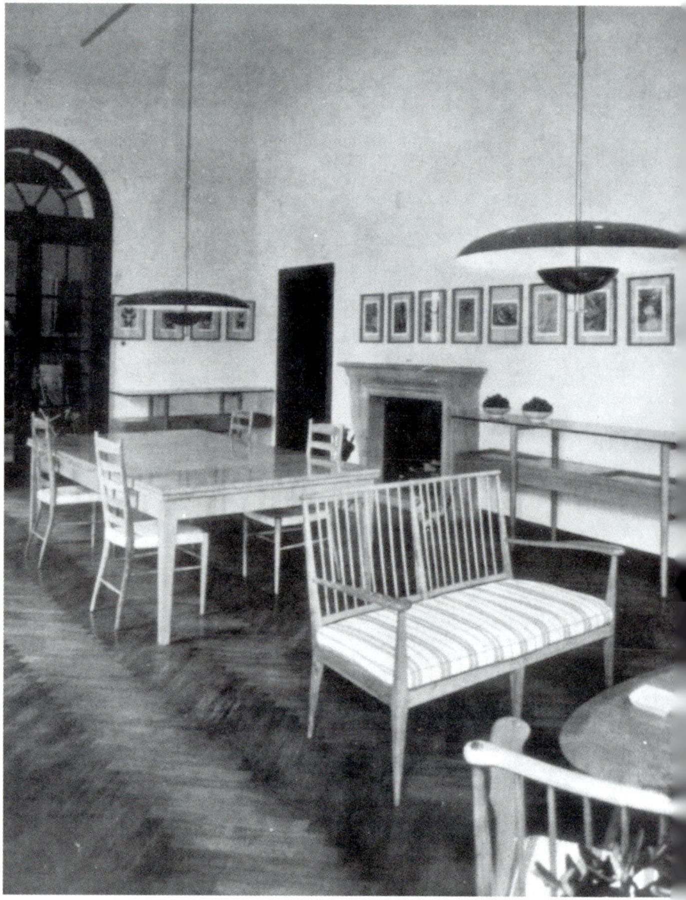

1941
furnishings for a house
in Bergamo
Private Collection

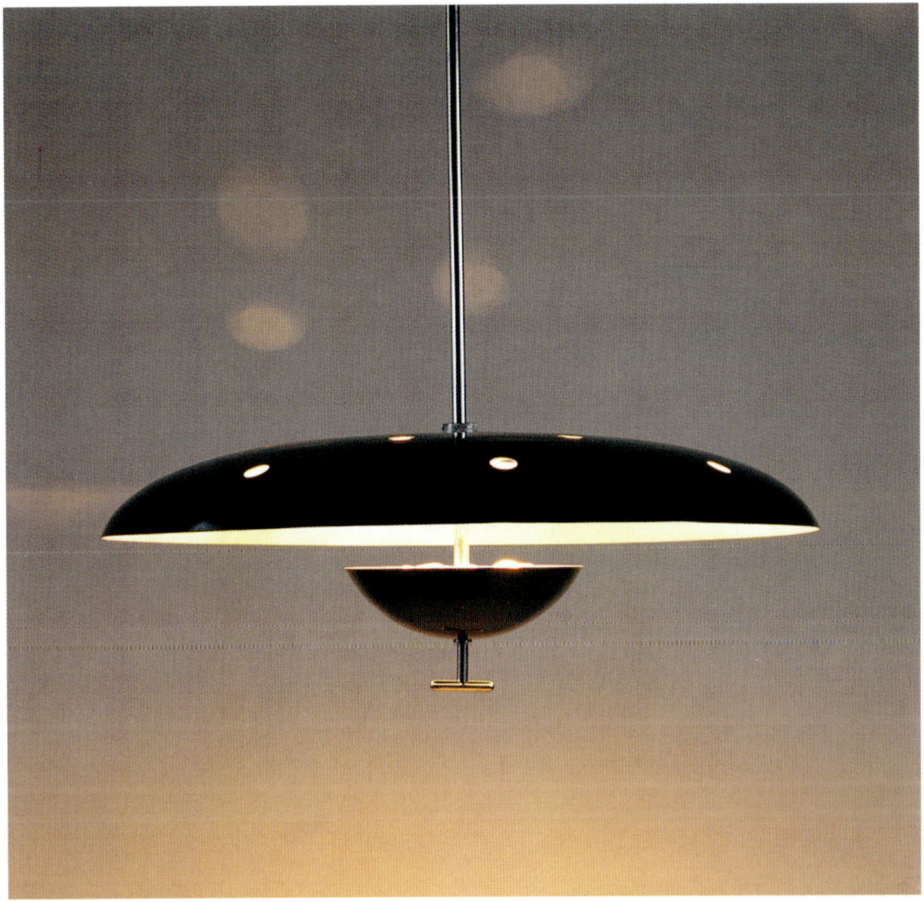

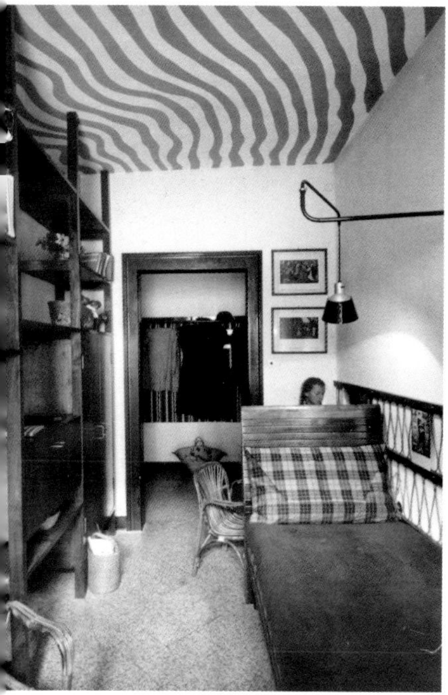

The children's room in the house in Bergamo.

"I am not an architect of exteriors. Even though my exteriors are not so ugly, I am primarily an interior designer. To solve furnishing problems I have at times had to invent furniture." This statement perfectly sums up Caccia Dominioni's relationship with furniture and interior design: the project always sprang up around a space, its intention being to make it more liveable and functional.
As is well known, Caccia Dominioni's journey in the world of design began in the years just before the outbreak of the Second World War, when the scarcity of work in the building sector led him to get involved between 1938 and 1940 in the creation of utensils, such as the *Miracoli* flatware, designed jointly with Livio and Pier Giacomo Castiglioni, and the five-valve radio receiver for Phonola. In the same period, before being drafted into the army at the start of the war, Caccia Dominioni created numerous furnishings for homes or even single rooms, including the pieces shown here, conceived for a house in Bergamo and published in trade magazines at the beginning of the forties.
Caccia Dominioni designed almost all elements ad hoc with great mastery and elegance, skillfully solving spatial problems. He managed, for example, to accommodate three girls in two rooms that could not be modified, one of which was very narrow, almost like the wagon of a train, with one uneven wall. For the dining room he instead designed a large

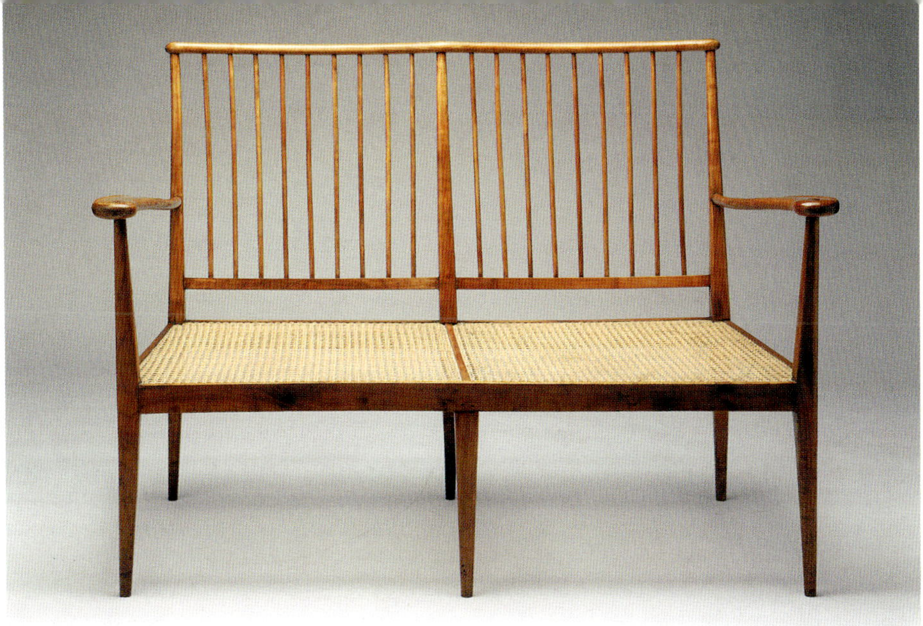

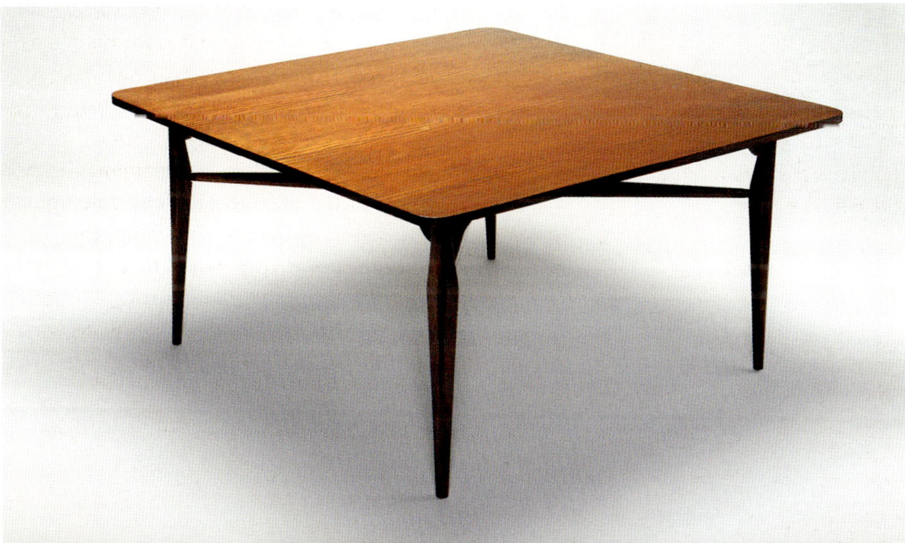

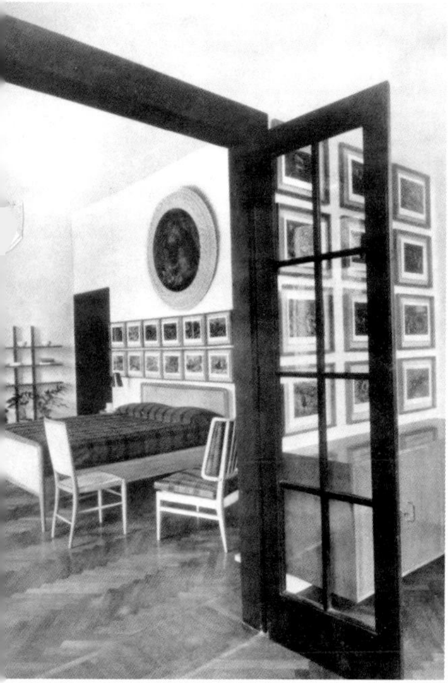

Bergère armchair in the bedroom.

extendible table that could seat up to eighteen people, surmounted by two hanging lamps, which could be moved to light it according to its length. He also designed a wing chair with a spacious low seat for the bedroom, while for the living room he contemplated a small two-seater sofa with the seat in pith, and a square table for the center of the room.
In one of the two children's rooms – characterized by furniture in polished cherry wood, wicker upholstery and the warm colors of green moleskin for the bed cover and black and yellow hemp for the large cushions – Caccia Dominioni introduced a lamp made of iron tubing that could be moved and directed. It was not designed ad hoc, but chosen from pieces already in production, foreshadowing a practice that would be established in subsequent years, but can already be glimpsed in its rudimentary stages in the choices of architects who included existing pieces like the *Tripolina*, *Sudanese* or *Chiavarina* in their projects.
All the furniture designed for the Bergamo house was made by famous craft workshops, such as Saita and Milani. The design precision and search for simplification of forms objectively anticipated the working method Caccia Dominioni would adopt beginning in the postwar period with Azucena, opting for small-scale production characterized by refined and sophisticated quality.

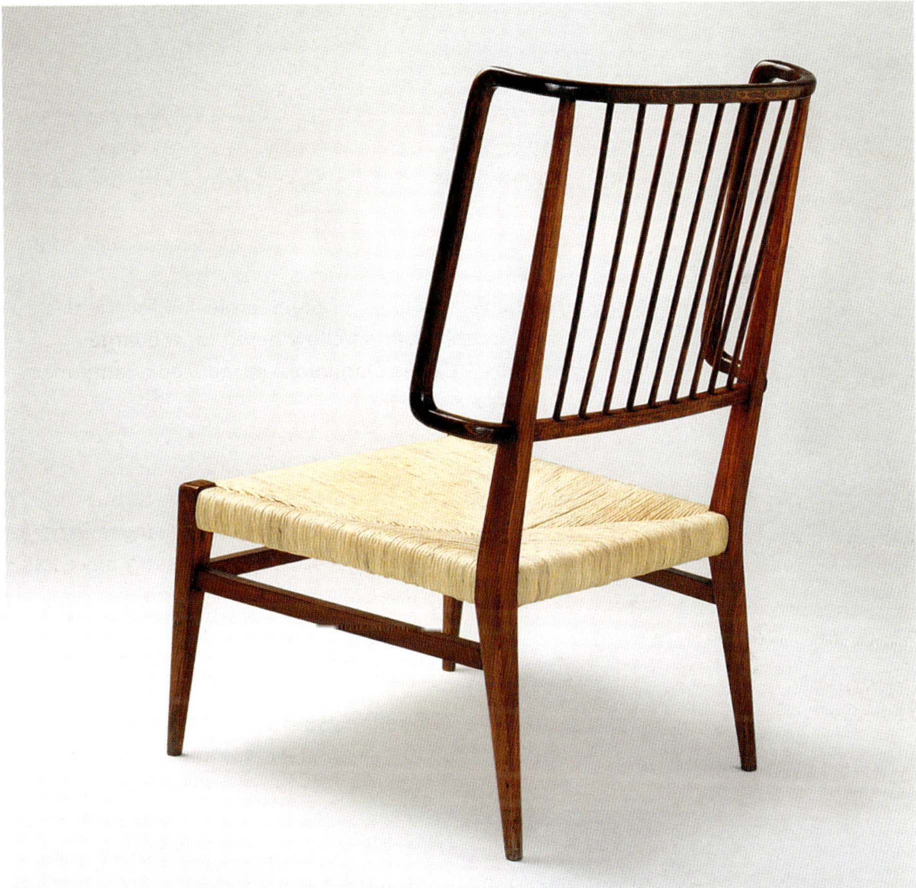

Pietro
Chiesa

In the early twenties Pietro Chiesa opened a famous artisan's workshop for the processing of glass and crystal, specialized in both new aesthetic and functional solutions and the experimentation of new technical manufacturing processes. On the basis of these experiences, "Chiesa entered Fontana Arte in 1933 as artistic director, commencing production that in the years between the two wars constituted a partnership between industry and craft that was as successful as it was rare." (*Rassegna,* no. 4 (1980), p. 56.)
The two models shown here originated from this partnership. The floor lamp is formed of a wooden stand with a twin series of adjustable metal cups, placed just below the top of the stand, the first three of which are white, while the last, the lowest ones, are made of brass. It is a prototype of the lamp made of metal with glass cups that Chiesa would show at the exhibition of decorative arts in Buenos Aires in 1938. The paired cups would also be used in a striking chandelier made by Fontana Arte in 1938, with a metal frame and diffusers positioned in decreasing numbers towards the bottom, producing a variation on the eighteenth-century Murano chandelier.

1936
floor lamp in brass and wood
Private Collection

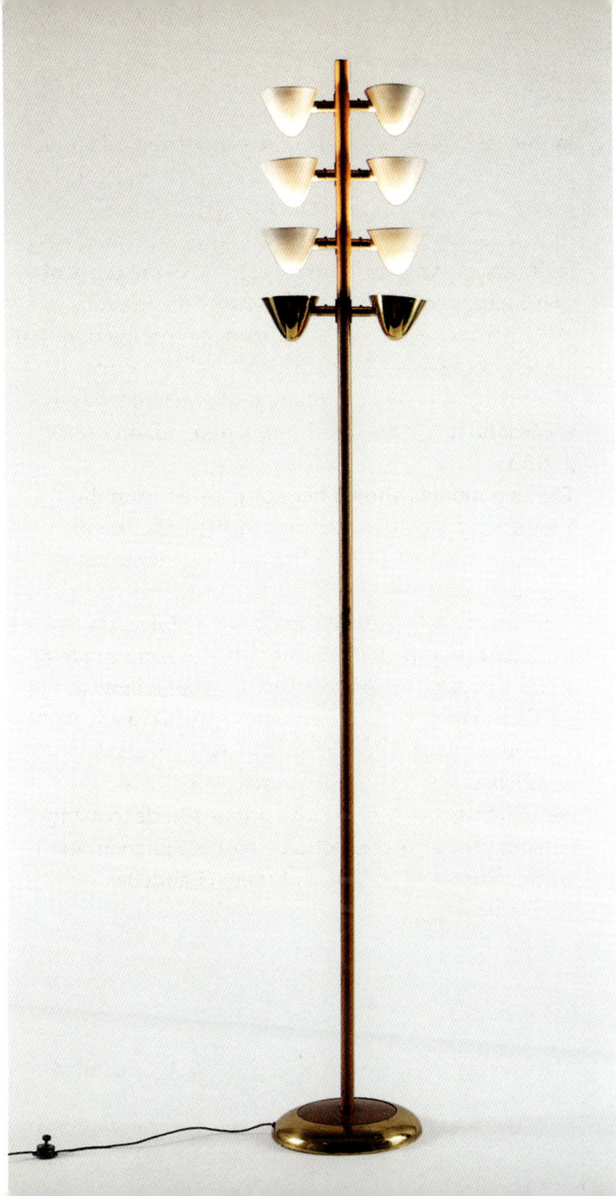

The table lamp is instead one of the variations Chiesa made on the same type: the base is crystal with greenish reflections, the lampshade is made of cloth. Chiesa worked a great deal on table lamps, alternating and varying both the bases (engraved mirrored glass, colored crystal and silvery metal with details in transparent crystal) and lampshades (white grainy paper, milk glass diffusers, and so on).

"Chiesa treats crystal like a precious stone. He grinds it, bends it and engraves it; he produces the tiniest pieces and monumental objects with equal solicitude. . . . His products are distinguished by the sophistication of his workmanship and the preciousness of the materials he uses; it is clearly luxurious and, despite being defined 'industrial,' essentially the product of craftsmanship: objects are produced in very limited series and items of furniture are almost always unique pieces." (Mauro Bonetti, in R. Bossaglia and V. Terraroli, ed., *Milano Déco. La fisionomia della città negli anni venti* (Milan: Skira, 1998), p. 194.)

1936
table lamp with lampshade in wood, crystal and brass
Private Collection

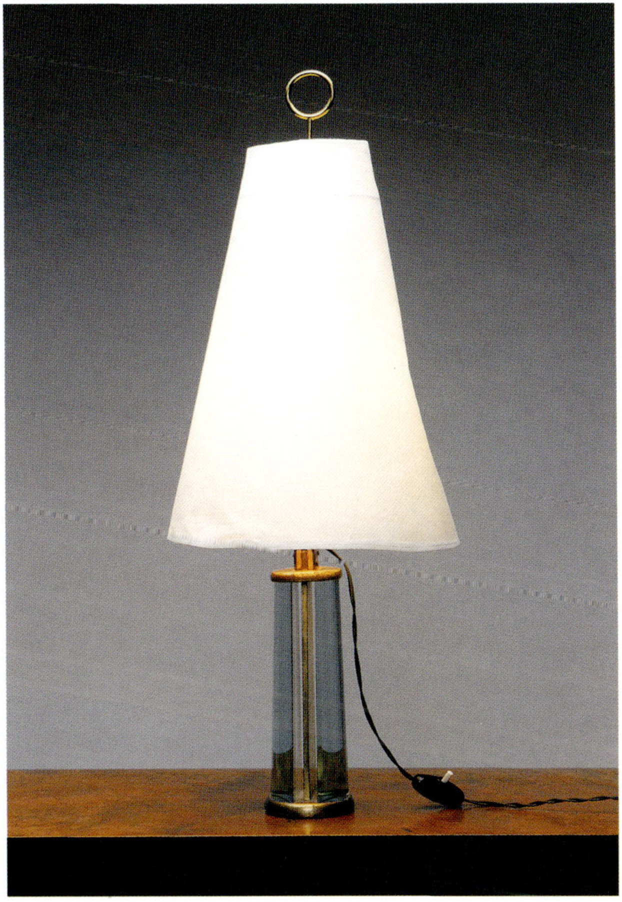

Luigi
Figini and Pollini
Gino

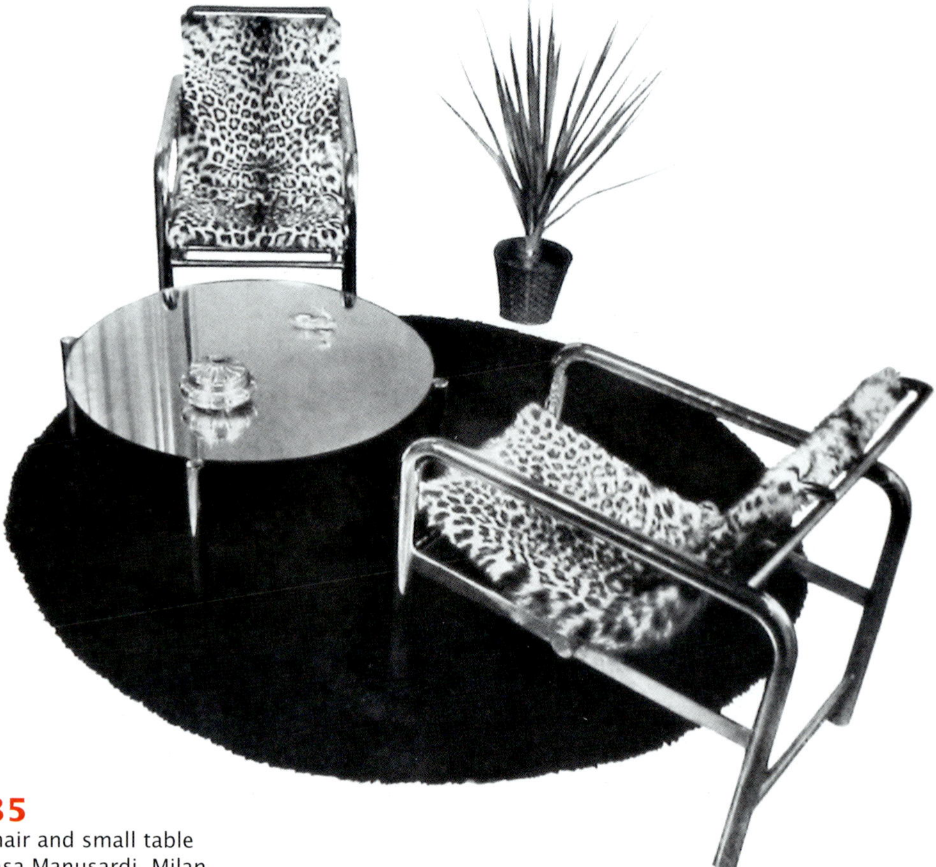

1935
armchair and small table
for Casa Manusardi, Milan
Private Collection, Milan

In Domus *15, no. 109 (January 1937)*

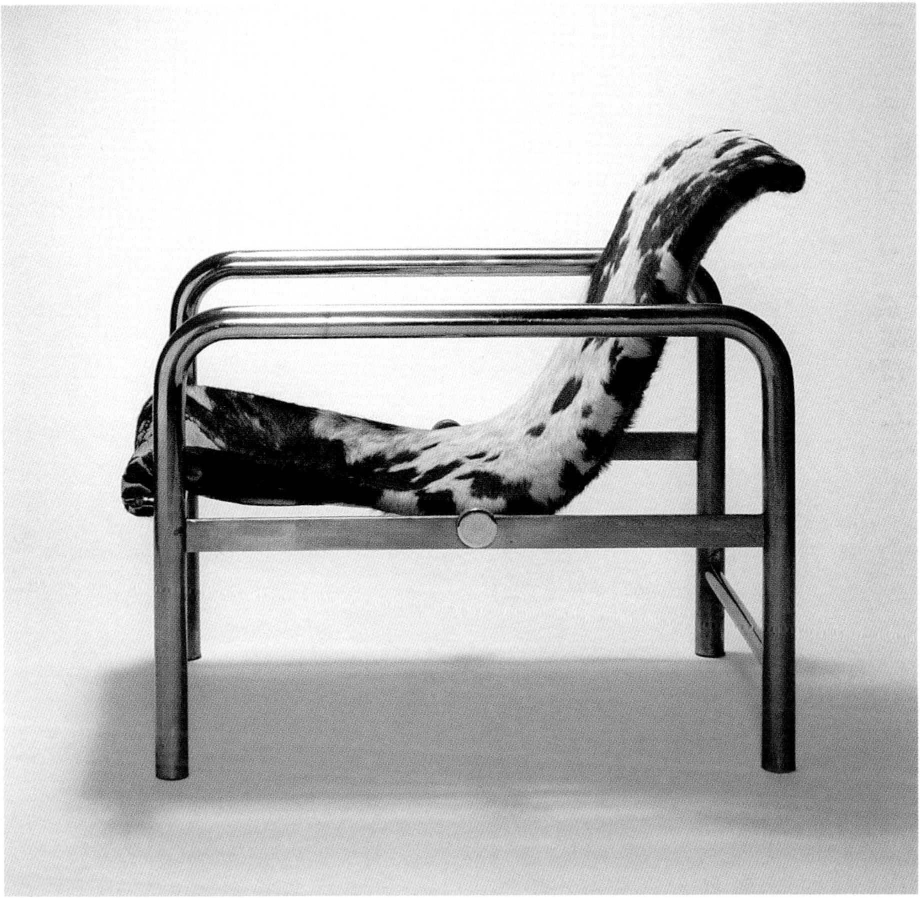

This small living room – composed of two armchairs and a small, round mirror-topped table – is part of the famous furnishings project conceived by Figini and Pollini in 1935 for the home of Doctor Gian Marco Manusardi, located on via Nirone, 11, in Milan. The materials were new. Metal and glass, brought in for their inexpensiveness and durability, were combined with the original Ethiopian leopard skin upholstery of the small armchairs, not just as a veiled tribute to the exoticism of imperial culture, but also to fuse the warmth and voluptuousness generated by the animal skin with the cold chrome steel of the tubular frame. The artificiality of the metal thus matched the naturalness of the upholstery according to a poetics of contrast, producing a deliberate and intriguing oxymoron. The original leather of the seat, completely worn-out and irredeemable, was subsequently replaced with cowhide, as seen in the model shown here.

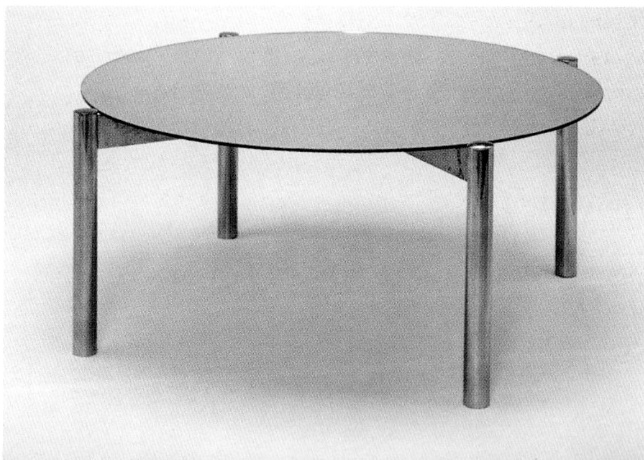

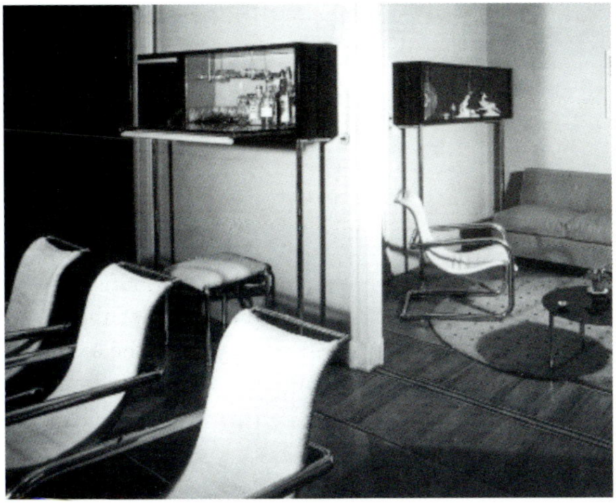

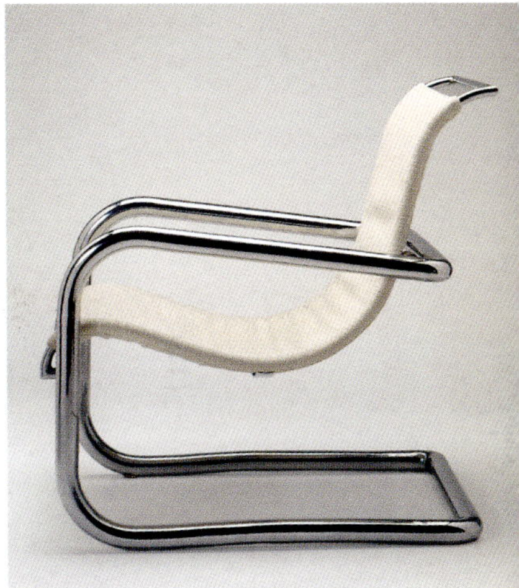

The seat upholstered in coarse canvas derives instead from the furnishings created by Figini and Pollini in 1935 for the Ferrari House in Milan. If in the small chairs of the Manusardi House the cross element is broken up by a milled washer, mounted like a precious stone in the frame of the furniture with the task of regulating the seat's carrying capacity, then in the Ferrario House's model the chrome tube is instead a cantilevered continuum, in compliance with a refined and extremely modern play of variations on the theme of the use of a specific material. Designed ad hoc as unique objects, these pieces of furniture were conceived even then in view of eventual industrial production.

1935
armchair
for Casa Ferrario, Milan
Private Collection, Milan

Ignazio
Gardella

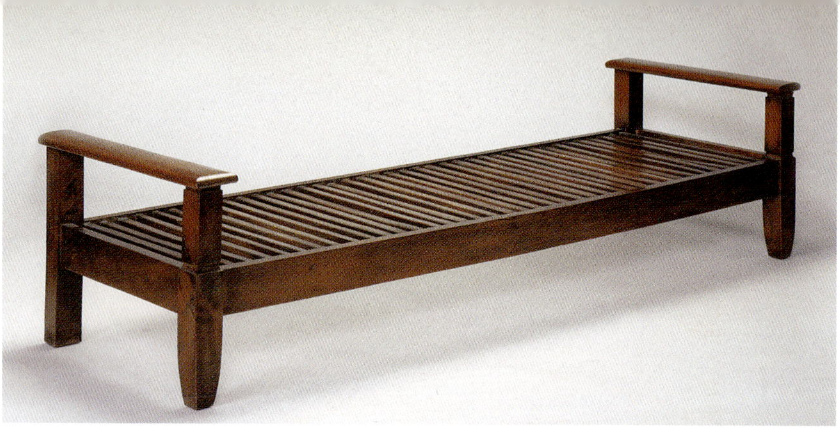

The armchair, constructed entirely of cherry wood, maintains absolutely balanced proportions: the depth and height of the backrest measure 80 cm each; the seat is made of wood strips measuring 2.5 cm x 2.5 cm, set 3 cm apart. The backrest is S-shaped and reclinable to facilitate transport, and the cushions are made of natural jute. The armchair is coordinated with a backless bench, also made of cherry wood and marked by narrow and slender proportions (length 210 cm, depth 68 cm, seat height 30 cm, cushion height 8 cm). Like the armchair, the bench seat is made up of wood strips with the same size and proportions.
Here Gardella tackled the theme of coordinating furniture conceived for the same environment at the design stage, thus facing one of the problems that mass production would have to solve on an industrial scale.

1932
living room
armchair and bench
Gardella Family Collection

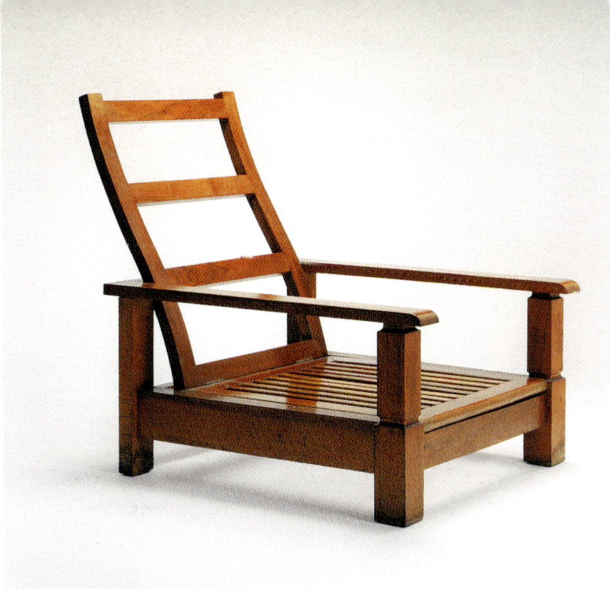

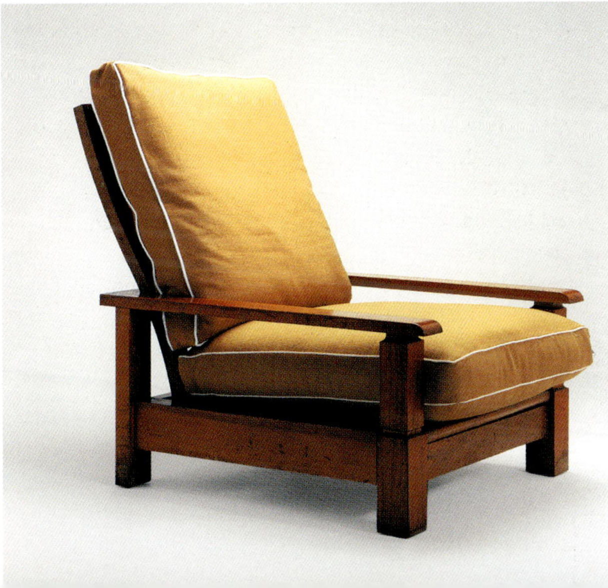

The lamp shown in the context of one of Gardella's interiors, in Domus, *no. 207 (March 1946).*

This floor lamp peeps out from between two *Tripolina* chairs in a living room shown on the pages of a 1940 issue of *Domus*. It is a prototype, or the forebear of a typology that Gardella would subsequently fine-tune with Azucena, but whose initial studies date back to the first half of the thirties. The frame is made of black fired metal measuring 3 cm in diameter, and features an extendible telescopic portion in chrome iron. The diffuser is a semi-spherical bell hood, measuring 20 cm in diameter, with a ring handle in chrome iron to facilitate the telescopic sliding. The height of the lamp and the angle of the stand are contrived to direct the light only onto the reading material held by the person sitting in the armchair.

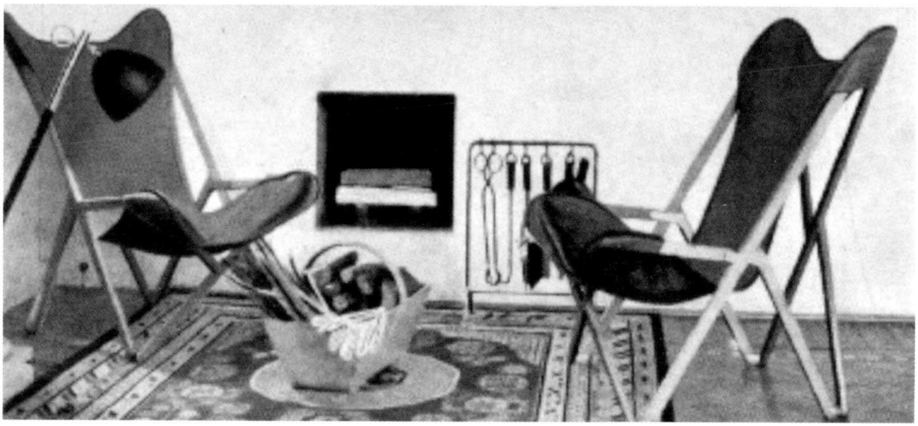

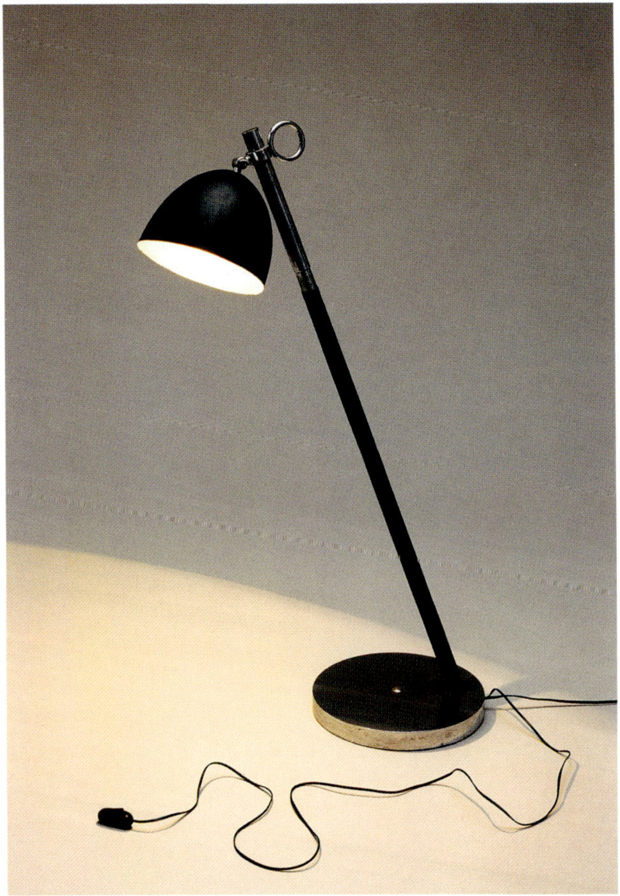

1934
floor lamp for armchair reading
Gardella Family Collection

Pietro **Lingeri**

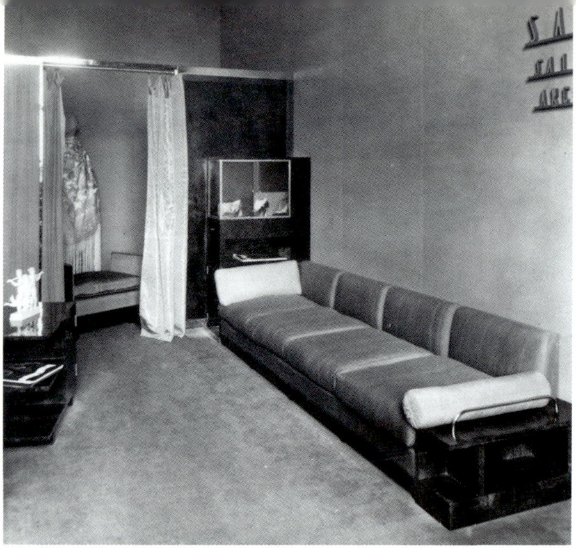

4th Triennial, 1930. "Dressmaker's shop, fitting room lounge." © Triennale di Milano Archives.

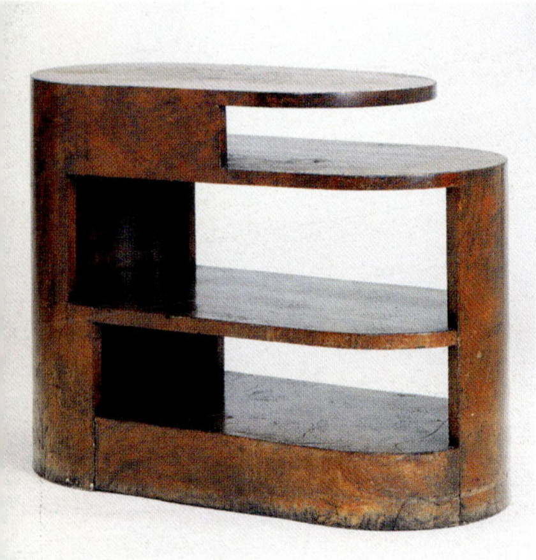

1930
shelf and cabinet for the "Fitting room lounge in the modern dressmaker's shop," for the 4th Triennial of Monza. Pietro Lingeri Archives, Milan

In 1930, at the 4th Triennial of Monza, a "Sartoria Moderna," or modern dressmaker's shop, was introduced and outfitted, composed of six rooms: Mario Cereghini's office for the manager, Gabriele Giussani's atelier, Giuseppe Terragni's entrance and model show room, Gianni Mantero's undergarments room and Pietro Lingeri's fitting room lounge. It was intriguing and fascinating not just because of the aesthetic and functional solutions it set forth, but also because of the "team" that produced it (in addition to the five designers, at least forty firms, including silk weavers, upholsterers, marble workers, tailors and lace workers representing the highest craft traditions of the Como and Brianza area, took part). One of its

best moments was Pietro Lingeri's fitting room lounge, composed of a series of three fitting rooms in walnut, lined with silk and mirrors, and "striped" with shiny metal bands. The furnishings included a sofa for the waiting area, small tables and the cabinets shown here, made by the Orsenigo di Figino Semenza firm. Ornella Selvafolta describes the largest: "It is a perfect parallelepiped structure, obeying strict size ratios, where the height is four times the width and twice the base (the original base was, in fact, replaced with taller supports in tubular metal). To this uncompromising and simple geometry, Lingeri added an asymmetric design and played on the contrasts of the materials, staggering the layout of the compartments, emptying and arranging the volume differently through the addition of a drawer, a pullout compartment on the side and corner display cases, juxtaposing the nakedness of glass and metal with the softness of the rich shot walnut." (G. Polin, O. Selvafolta, *Mobili come aforismi. Trentacinque mobili del razionalismo italiano* (Milan: Salone del Mobile-Electa, 1988.)
The smaller piece also features a walnut finish, and plays on shadows and depths through the gaps that break up the alignment of the various shelves all the way up to the cantilevered shorter top, which escapes, at least in part, the constraints of the structure.

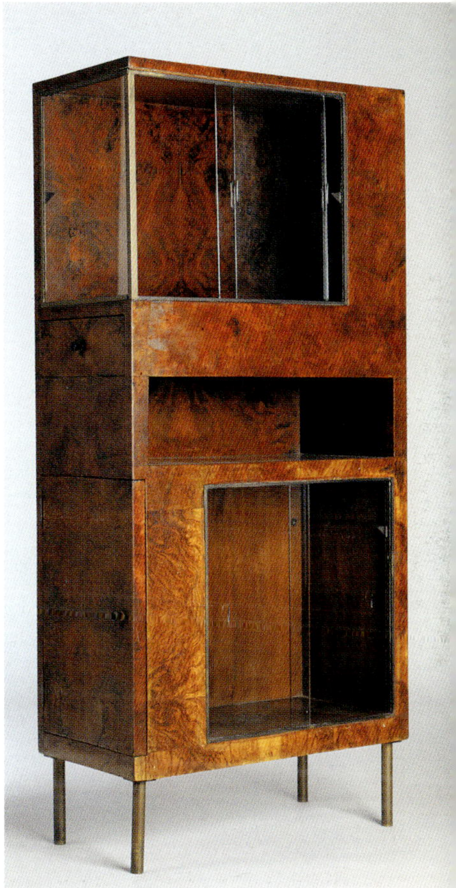

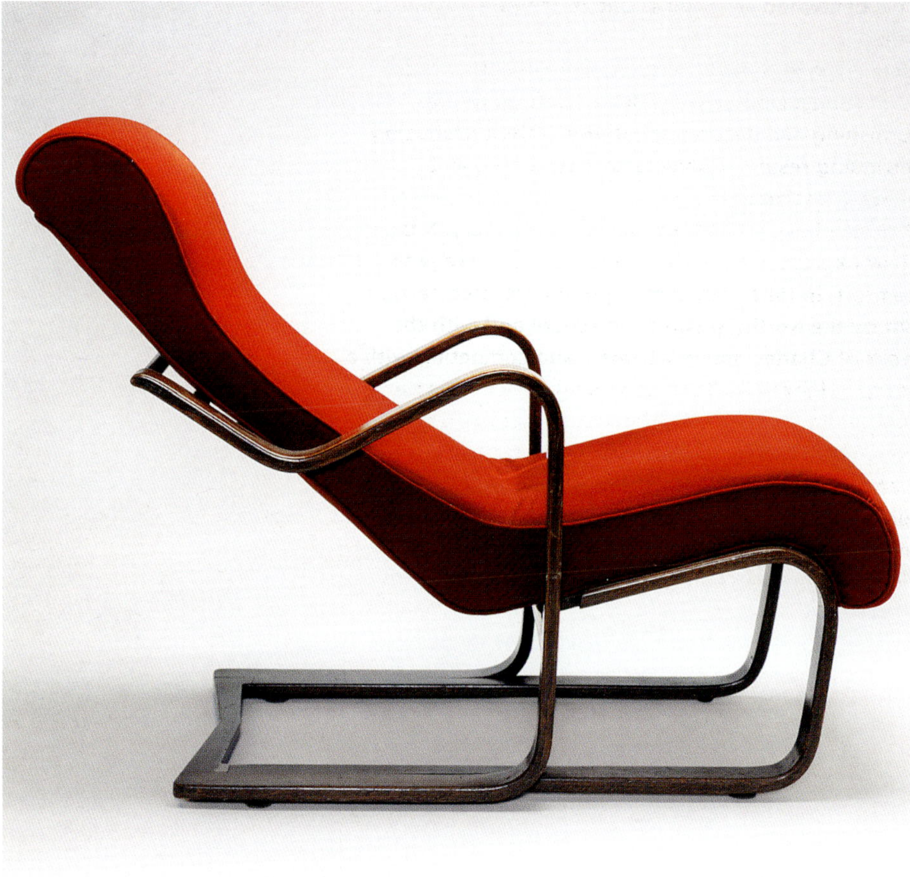

This bentwood armchair is one of the rare examples of a single piece of furniture made before the war by Pietro Lingeri. More interested in architecture than furnishings, Lingeri occasionally took on the task of furnishing and decorating interiors, often producing intriguing results. Characterized by a fluid and sinuous design with several curved ribs that depart from the base to support the soft upholstery, this chair expresses an intrinsic vocation for bourgeois comfort, in tune with contemporary research carried out by the Northern European school and with the work of Charles and Ray Eames, and competing with a stark and formally flawless structure. According to the information provided by the archive containing the documents and drawings of architect Pietro Lingeri, it was conceived to furnish Villa Silvestri at Tremezzo, built in 1930 in the district of Como. The piece was originally upholstered in olive-green Lenci cloth.

chaise longue
n.d.
Pietro Lingeri Archives, Milan

Giuseppe
Pagano

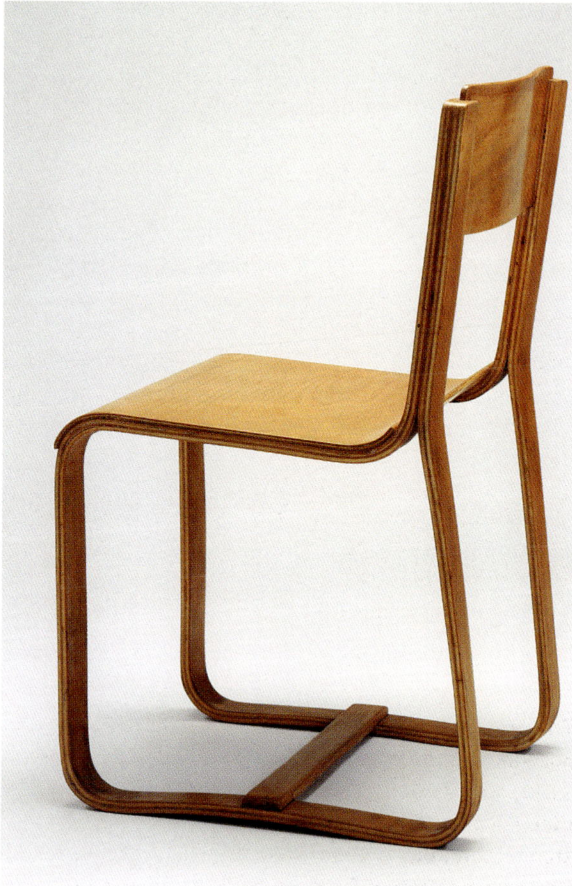

1942
chair for the canteen of
Università Commerciale
Bocconi, Milan
Alessandro Pedretti Collection

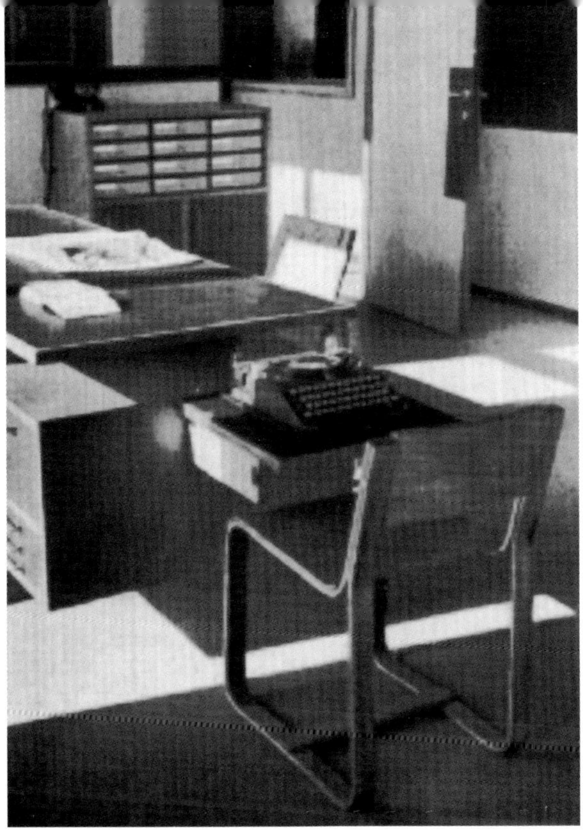

The chair in the administration's secretarial office, Università Bocconi, 1942, in Domus *20, no. 170 (February 1942).*

The project for Milan's Università Commerciale Bocconi, commissioned in 1937 and inaugurated in 1942, required that Giuseppe Pagano – and partner Giangiacomo Predaval – not only design the building's architecture, but also its interiors and furnishings. Drawing inspiration from his beloved Alvar Aalto –

Corridor between the chancellor's and president's offices, in Domus *20, no. 170 (February 1942).*

who between 1929 and 1933 had been assigned the task of designing the interior furnishings of the sanatorium he was building near Paimio, creating ad hoc a chair in curved plywood that soon was produced on a large scale – Pagano also worked on commission, filling a specific order. He attempted, however, to put into practice an idea of modularity and standardization with individual items of furniture. For example, he thought of the chair as "a single loop of curved beech" to which "different backrests could be added in order to make different types of chairs." (Alberto Bassi and Laura Castagno, *Giuseppe Pagano* (Rome-Bari: Editori Laterza, 1994), p. 124.) The same design also generated, with modest variations, the armchair produced by the Maggioni Company used to furnish Casa Banfi in 1939. (*Rassegna*, no. 4 (October 1980), p. 85.)
The model shown here, with its orthopedic shape, was conceived for the canteen, but also for the bursar's or secretary's offices, and was designed so that it could be pulled up close to the typewriter. The tone is spare, essential, plainspoken: Pagano sought very simple symmetries that could create living spaces and functional objects, dictated by that "proud modesty" that is the moral cornerstone of modern architecture.

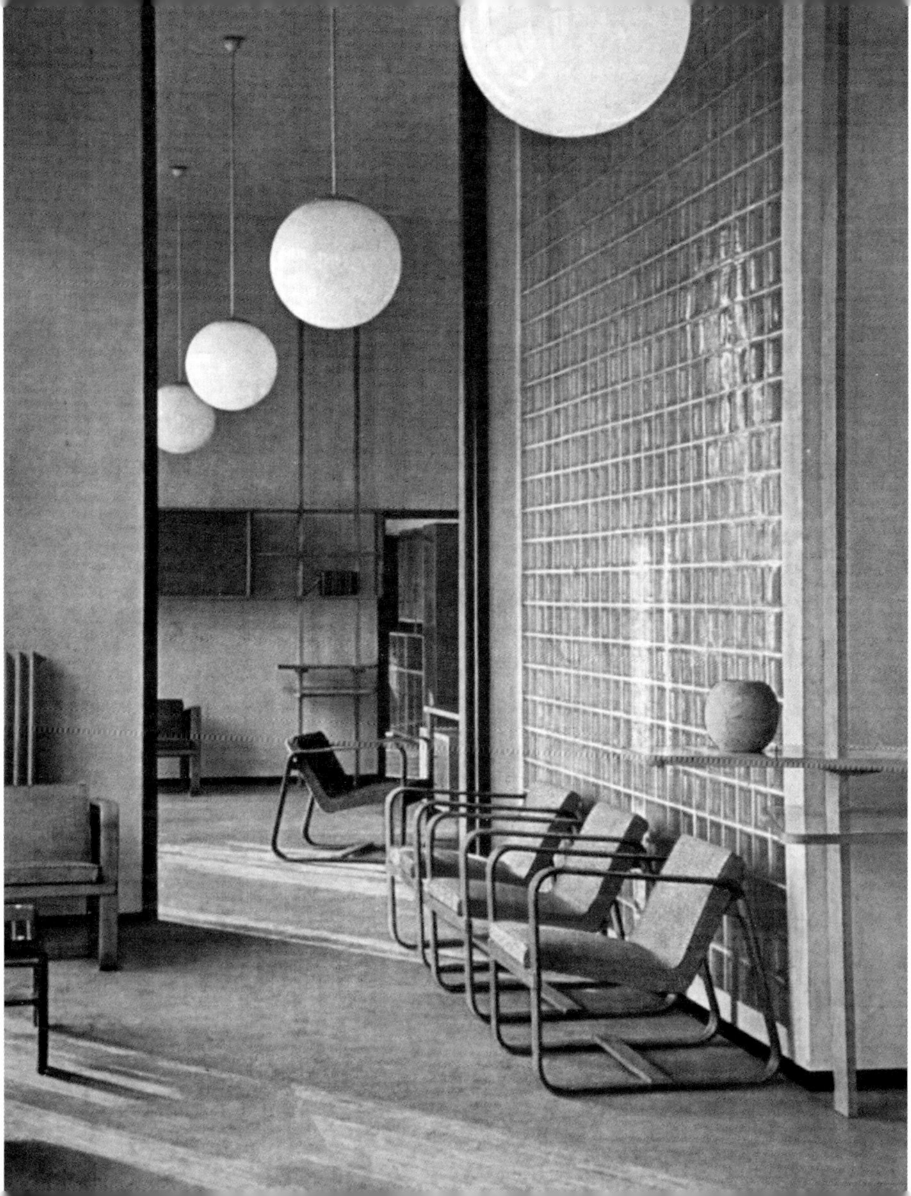

Gio **Ponti**

Aluminum chairs for operators at the central switchboard, Montecatini Building, in Casabella Costruzioni, *no. 138-139-140 (June-August 1939).*

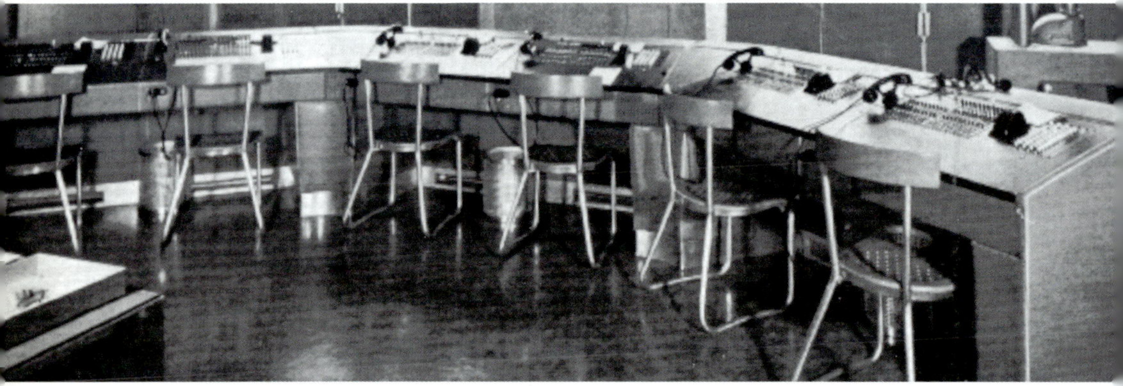

1938

chairs
for Montecatini Building on
via della Moscova, Milan
Alessandro Pedretti Collection

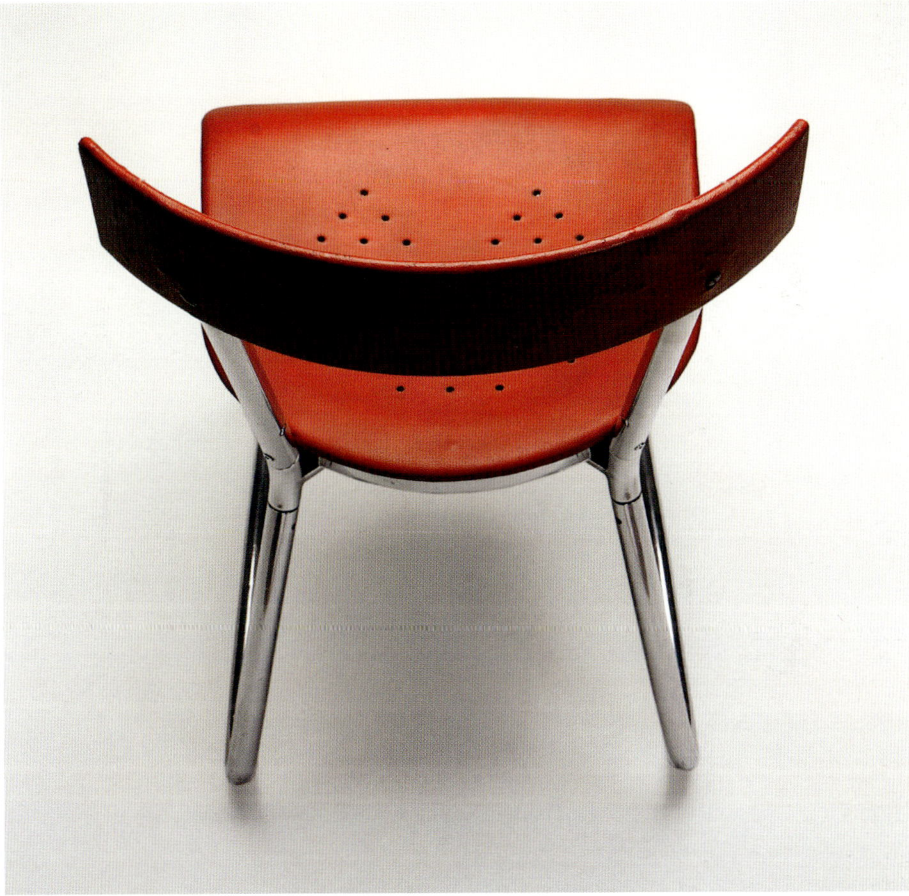

On the following page, from left, aluminum chair for office workers (Kardex), aluminum swiveling chair for typists (Kardex), aluminum armchair for executives (P. A.S.), and aluminum armchair for visitors (P.A.S.), in Casabella Costruzioni, *no. 138-139-140 (June-August 1939).*

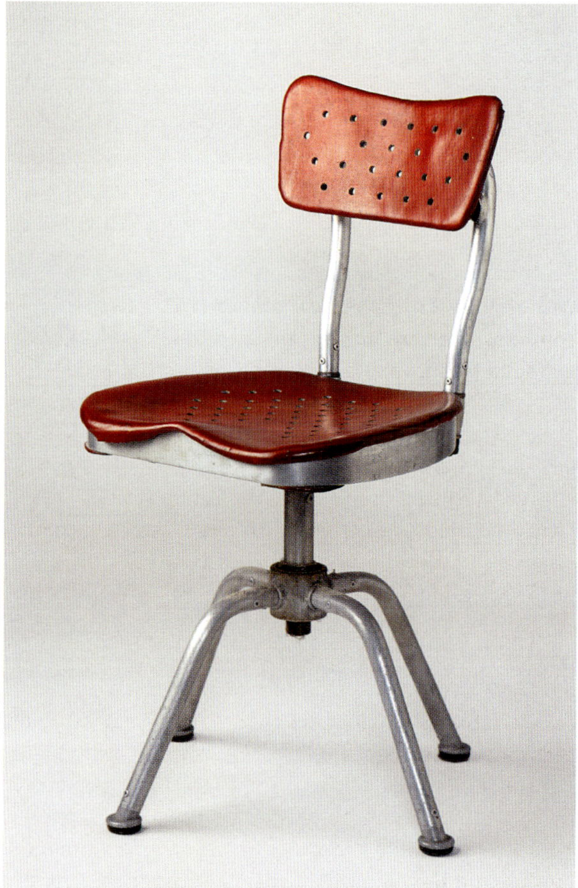

Private Collection

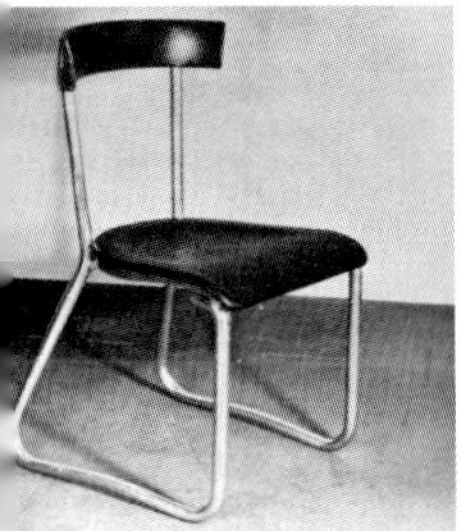
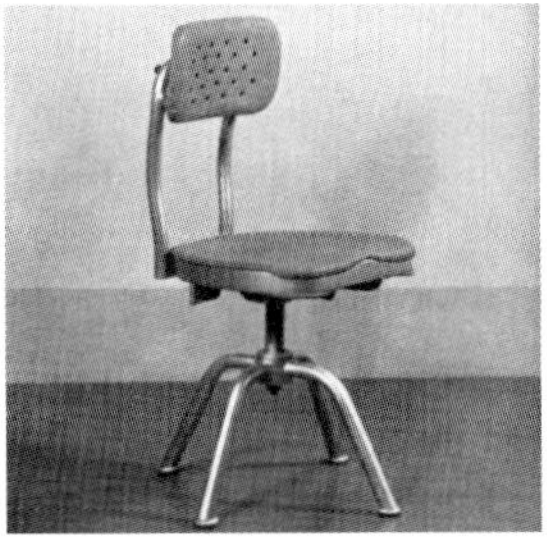
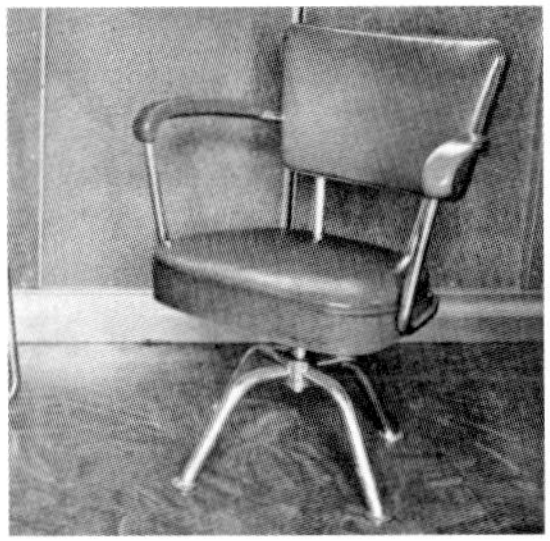
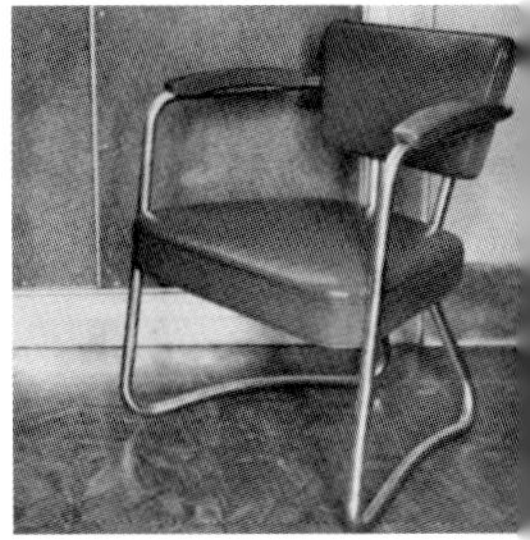

With the design of the new offices (1936-1938) commissioned by Montecatini and built in Milan between via della Moscova and Largo Donegani, Gio Ponti provided one of Italy's first and most important examples of "integrated design." The architect not only attended to the balance of the architectonic composition and critically studied the entire plan, but also dealt with the functional design of the furnishings to complete the offices.

For the seating area, for example, Ponti designed a sampling of aluminum chairs based on the principle of variation on a theme: a typist's chair, a swiveling chair for executives, an armchair for visitors and a comfortable and functional chair for secretaries. Ponti did not modify the manufacturing procedures typical of iron furniture adopted in the heyday of companies such as Parma of Saronno and Kardex of Milan, which held the contracts for the job. Instead he created new molds for cutting and bending metal sheets in order to diversify

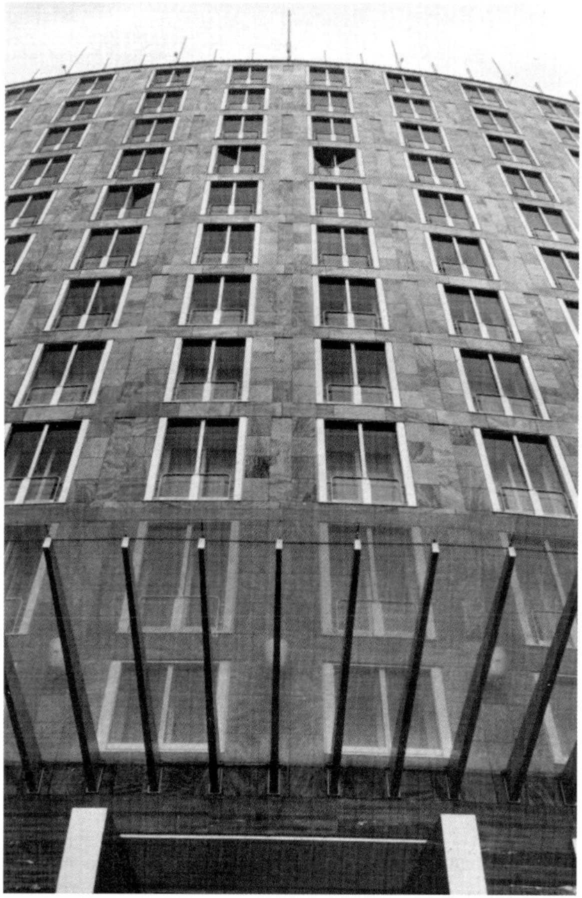

Montecatini Building, view of the central structure, Gio Ponti, Eugenio Soncini and Antonio Fornaroli, in Casabella Costruzioni, *no. 138-139-140 (June-August 1939).*

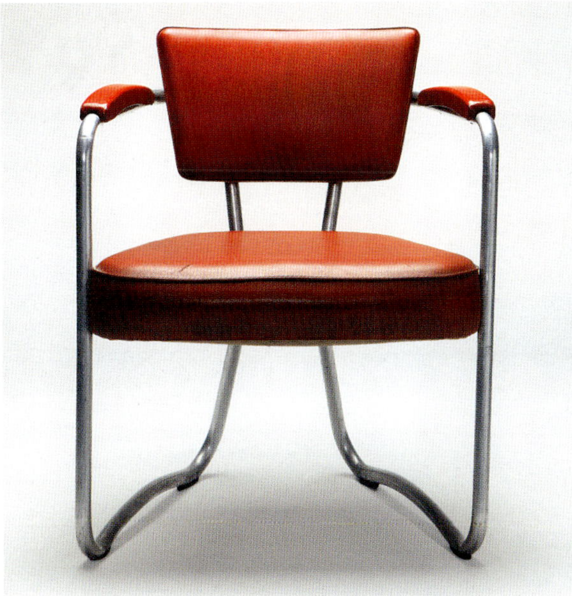

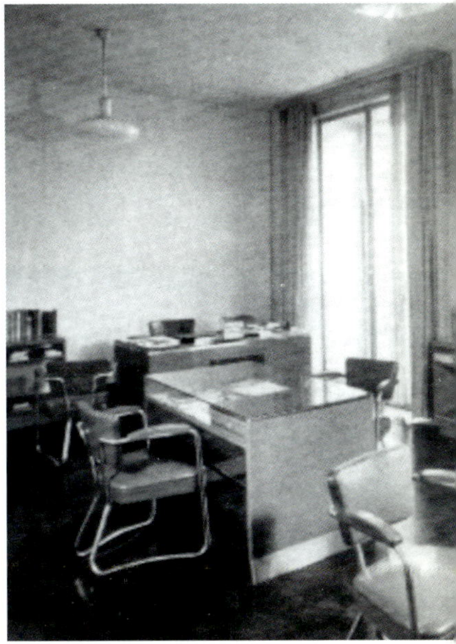

Metal furniture used in the building's wings, in Casabella Costruzioni, *no. 138-139-140 (June-August 1939).*

and distinguish individual models, conserving and optimizing the merits of a purely and typically productive approach. The manufacturing phases of these items of furniture, analytically described in *Casabella* 11, no. 138-139-140 (June-August 1939), pp. 133 ff., provide a testimony of how these productions were indeed a pioneering and advanced pilot experience in the process that would lead to the success of mass-production culture in Italy.

Private Collection

Piero
Portaluppi

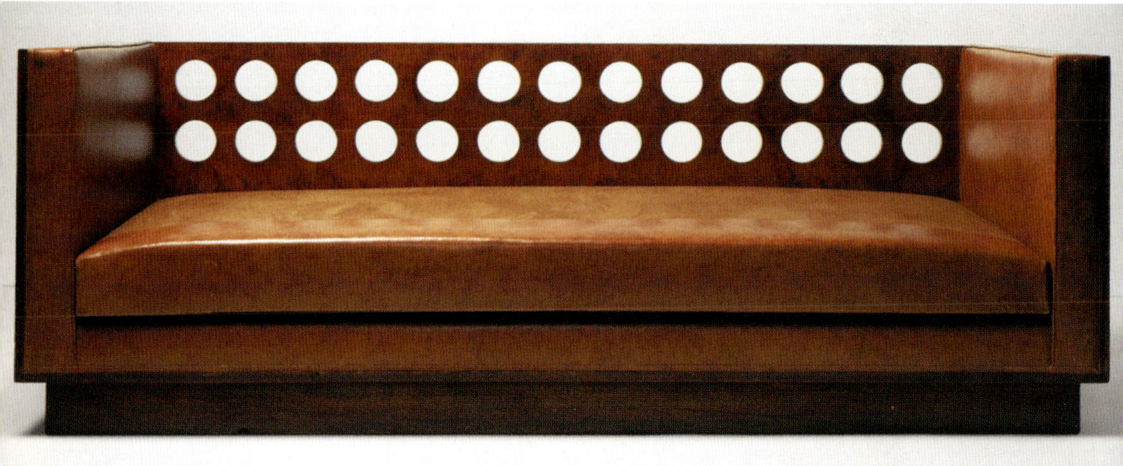

1938
sofa
for apartment on via Manin 35, Milan
Private Collection

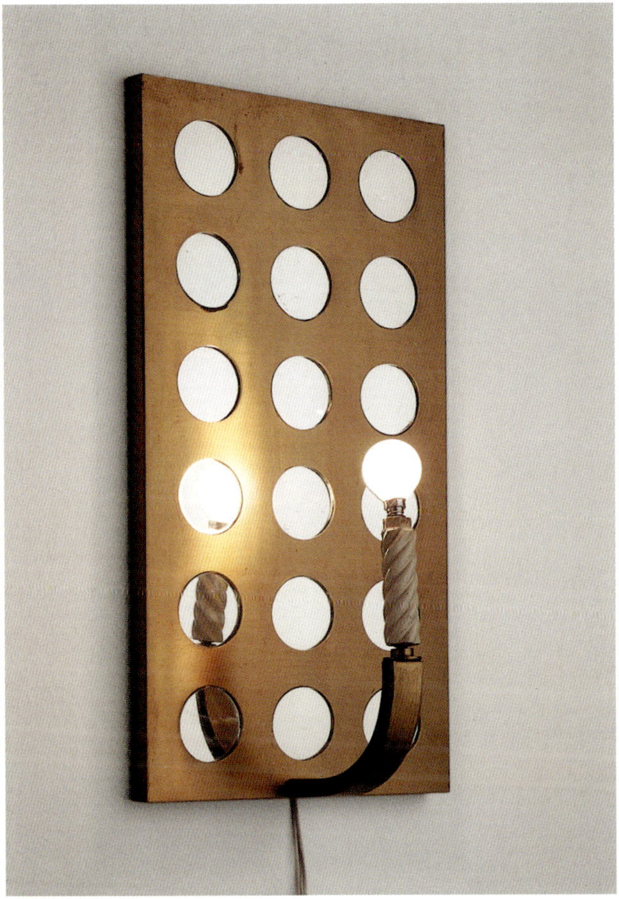

1934-1937

lamp, console, tabourets for Casa Corbellini-Wasserman, viale Lombardia 17, Milan
Private Collection

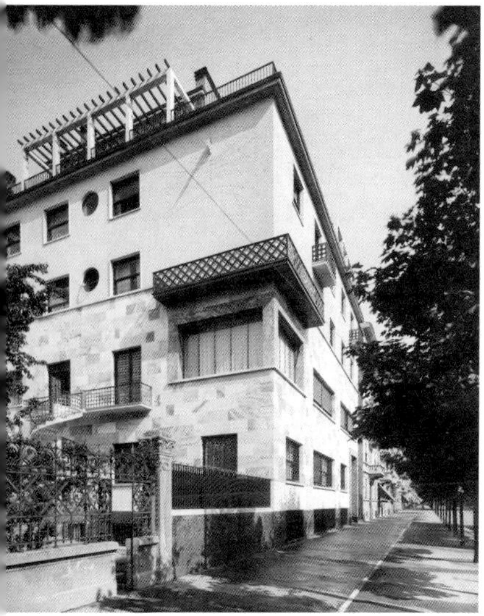

Casa Corbellini-Wassermann on viale Lombardia 17, (1934-1937). The house seen from the street. *© Fondazione Piero Portaluppi Archives, Milan.*

Private Collection

The eclectic relationship with the modern that characterizes Piero Portaluppi's architectural projects can also be found in the "minor" episodes that saw the Milanese architect engaged on the furniture and furnishings design front. From this point of view, one of his most significant creations was the house built for the Wassermans on viale Lombardia 17, located close to a pharmaceutical factory. Inaugurated in 1937, it is a small building with articulated volumes, a kind of "urban villa" with a facade featuring portholes on the upper two floors, and made of stucco to Portaluppi's specifications by the decorator Fontana. In this case the architect also designed the furniture: a tabouret with Doric columns and three circular tops, a low table with inlays in semiprecious stones and a "honeycomb" base, a console in briarwood of blatant neo-classical taste with metaphysical echoes, *quasi* De Chirico, and two brass bracket lamps with mirror circles and a light source bereft of reflectors or hoods. Innovation and tradition mingle in a deliberate blend of tastes, where the unique piece featured elements that were a prelude to production on a larger scale. The other piece shown here is a long deep sofa with a continuous geometrical structure, designed by Portaluppi in 1938 for a private dwelling on via Manin in Milan. Here the element of major interest is the backrest worked with a motif of regular and fretted circles, almost as though to confer something

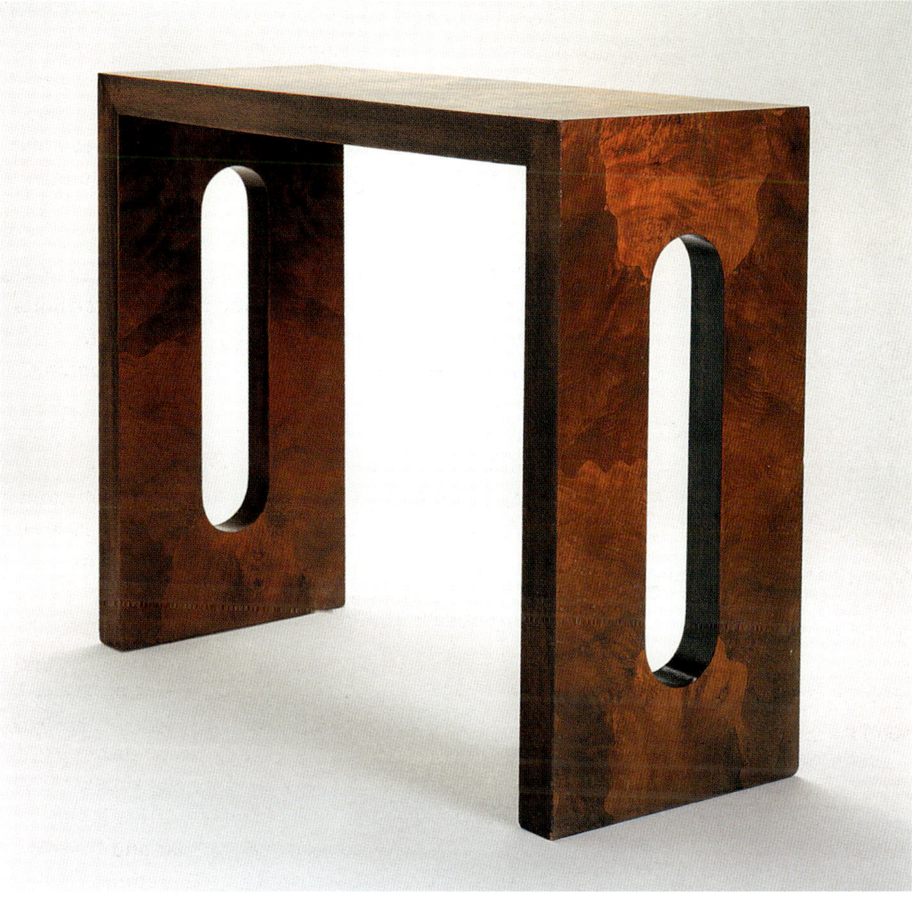

transparent and metaphysical onto a structure which is per se excessively "monumental."

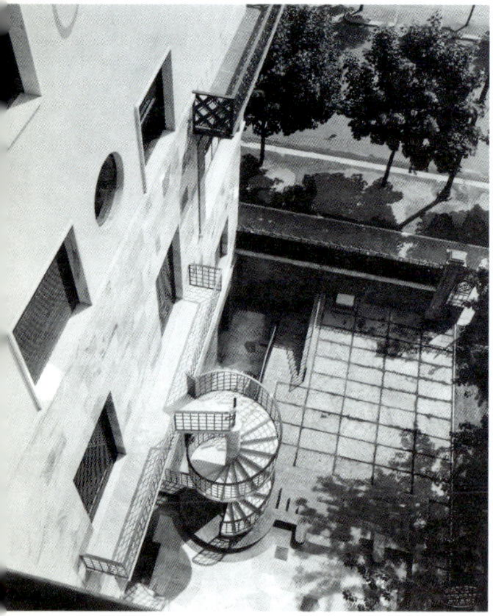

Casa Corbellini-Wassermann on viale Lombardia, 17 (1934-1937). The external spiral staircase connecting the two master apartments and the garden.
© Fondazione Piero Portaluppi Archives, Milan.

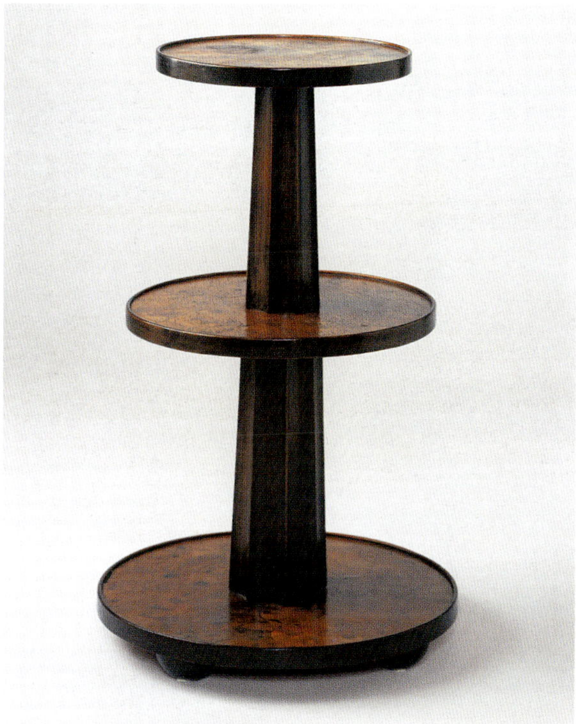

Private Collection

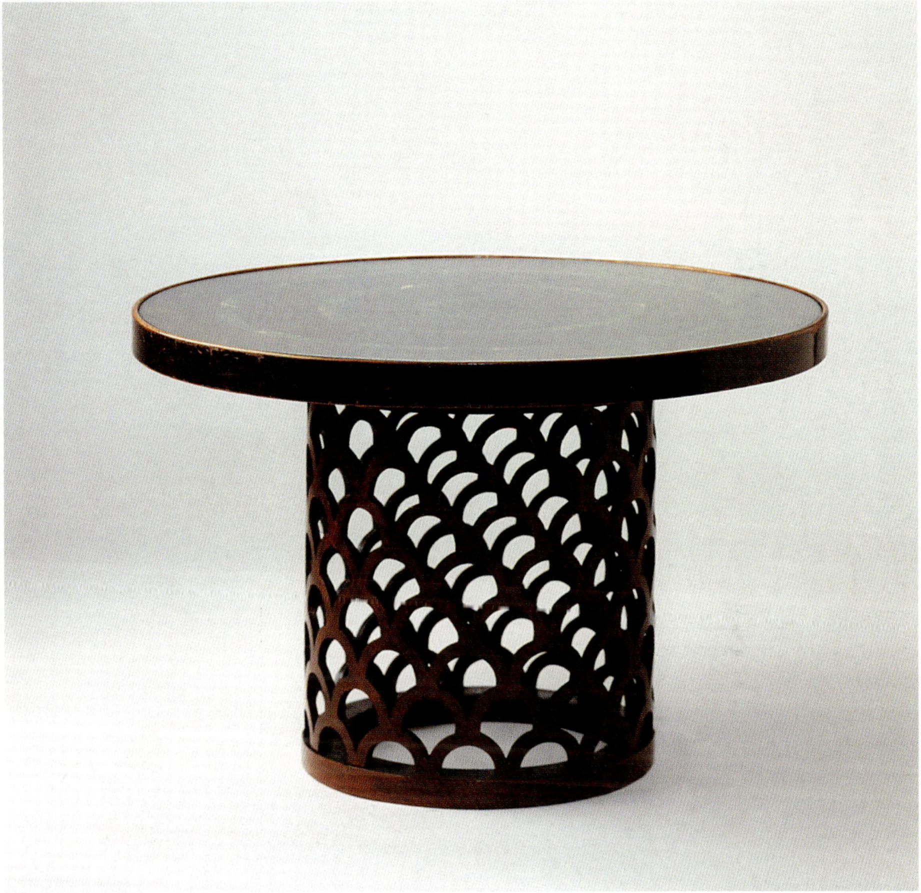

Gustavo
Pulitzer

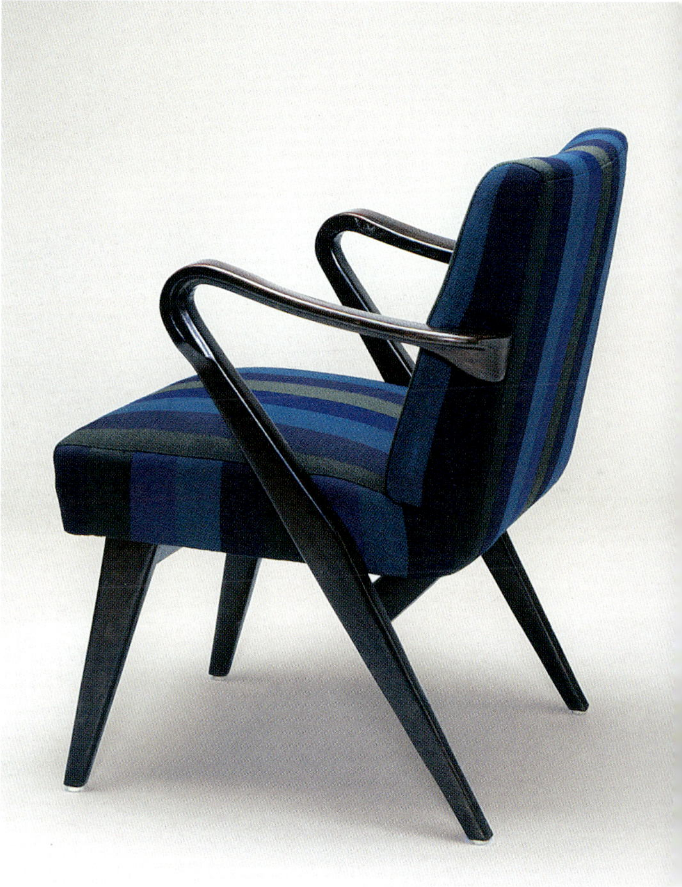

Interior of the cruise ship Andrea Doria, furnished with chairs, armchairs and furniture produced by Cassina.

1945

chair Mod. 526
for the Andrea Doria
produced between 1945 and
1953 by Cassina
Cassina Collection, Meda

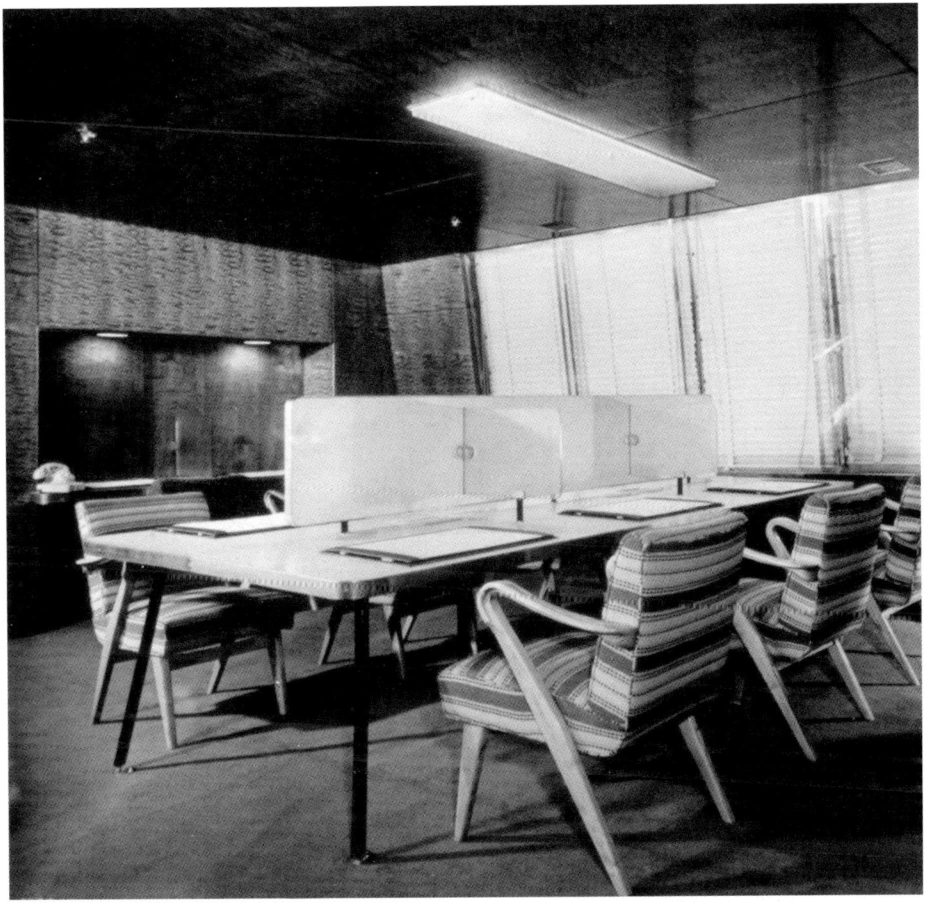

Designed in 1945 and put into production between 1945 and 1953, Pulitzer's chair has a walnut frame, with seat and backrest upholstered with blue and green striped fabric. It is an emblematic example of furniture conceived for ship fittings, which from the forties onwards represent one of the more emphatic spurs towards mass production. Irene de Guttry and Maria Paola Maino write: "Between 1948 and 1958 the fate of the company [Cassina] was decided by a series of orders for ship fittings (the largest orders would be those for the Conte Grande and the Andrea Doria), so large as to bring about profound changes within the company in terms of reorganizing work and purchasing new machinery. It was a question of producing an elevated number of pieces of particularly sturdy armchairs with stiff upholstery in fabric or leatherette, with splayed wooden legs, designed by Gio Ponti, Nino Zoncada and Gustavo Pulitzer." (Irene de Guttry and

Cruise ship Andrea Doria

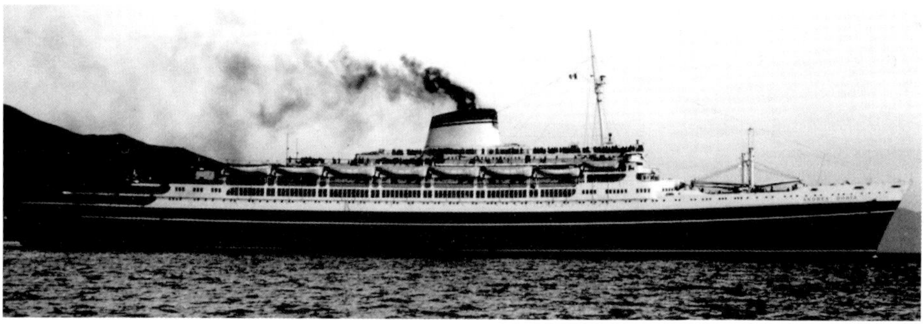

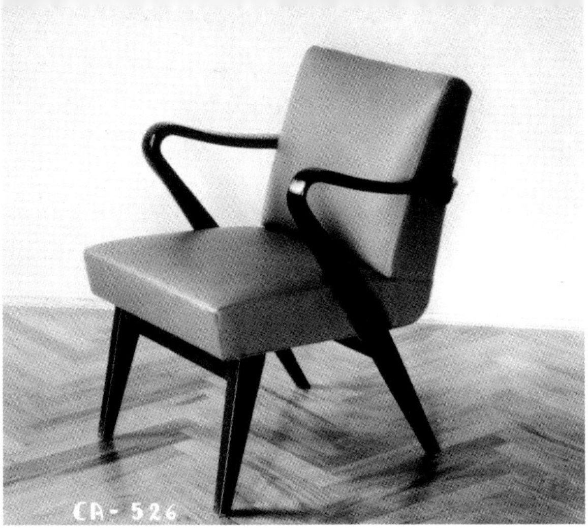

Maria Paola Maino, *Il mobile italiano degli anni ’40 e ’50* (Rome-Bari: Editori Laterza, 1992), p. 130.)
There is no lack of critics, however, who have played down the scope for experimentation that these orders allowed from a historiographical perspective: “The most fashionable architects and famous artists were called upon to furnish these extraordinary floating palaces. They were forced, however, by the anachronistic monumentality of the order to hide behind sporadic formal lyricism, forfeiting the more advanced experimentation that could be seen in the contemporary work that occupied them in the areas of homes and living.”(Enrico Morteo, “L’abitacolo da trasporto,” in Manolo De Giorgi, ed., *45,63. Un museo del disegno industriale in Italia* (Milan: Editrice Abitare Segesta, 1995), p. 54.)

Guglielmo
Ulrich

Interior of the Galtrucco Shop in Piazza Duomo, Milan, checkout area.

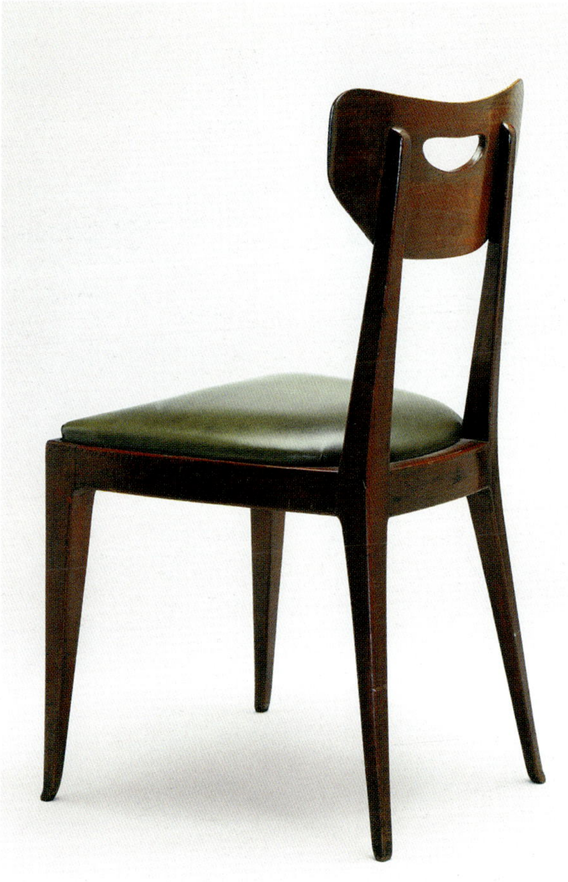

1949
chair
for the Galtrucco Shop in
Piazza Duomo
Lorenzo Galtrucco S.P.A.
Collection

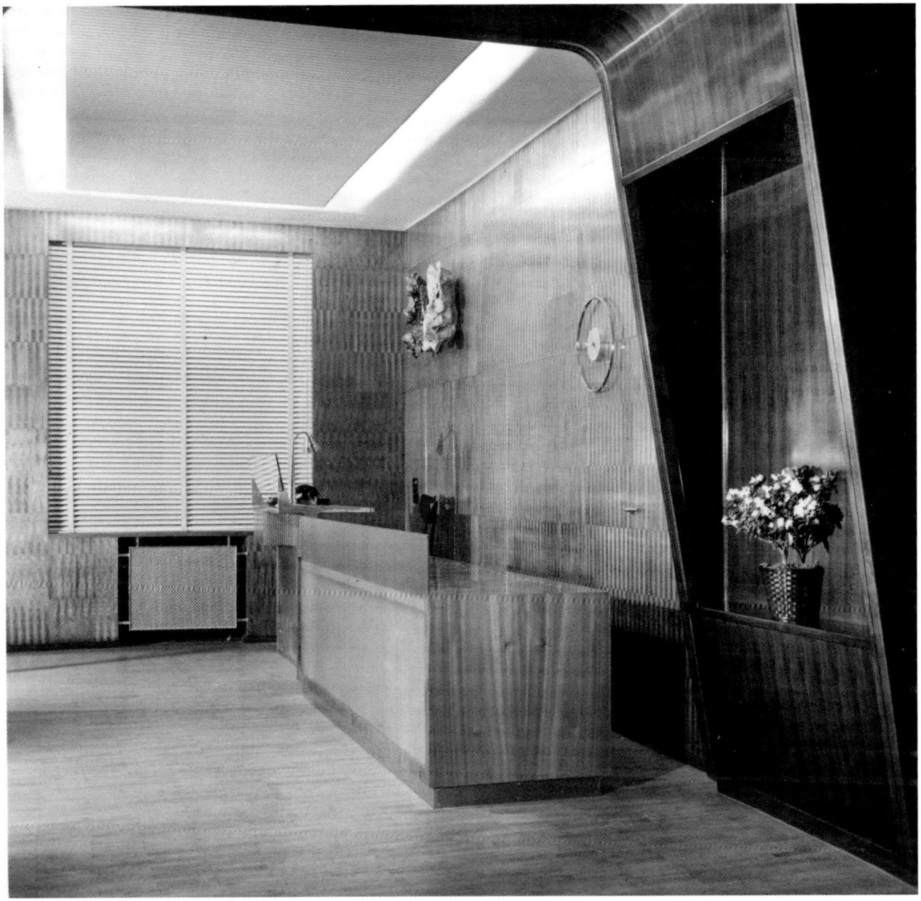

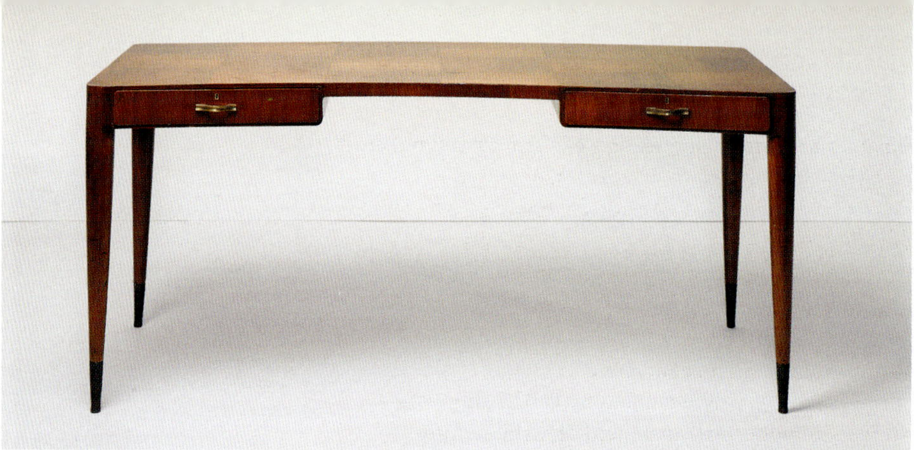

These three prototypes perfectly sum up the controversial relationship that an architect like Ulrich had with furniture design. While some of his pieces represent – as Melchiorre Bega wrote – "a true product of modernity," others must be considered more as beautiful variations on older models than "creations that fit well into present-day architectonic space." As such they often prove to be more interesting "as a sign of the value of the creator, than as part of the labored progress of modern art." (Melchiorre Bega, "Sedie di Ulrich," *Domus* 21 (December 1942), pp. 522 ff.).
Two of these three pieces – the pear wood armchair designed in 1937 by Alberto Pirelli and the cherry wood table with brass ferrules, feet and handles, which Ulrich himself designed for his own home in 1940 – reveal the very distinctive signature of a designer who was heir to the most sophisticated Italian twentieth-century tradition, while retaining an

1944
table
for Casa Ulrich
Giancorrado Ulrich Collection

obvious and proud suggestion of their prototypal craft natures. On the other hand, the chair that Ulrich designed at the end of the forties for the Galtrucco Shop in Piazza del Duomo in Milan, is more interesting in terms of the development of mass-production culture in Italy. Based on the search for a style determined by the combined presence of elegance and functionality, every detail of the shop furnishings – jointly designed by Ulrich and Bega – suggests "modernist" taste and stylistic features. Ulrich's chair was specially conceived for the checkout area, which has white stucco ceilings and fluorescent lighting, walls striped alternately in teak and satin walnut, and flooring in dark travertine and striped green rubber. The chair's upholstery picks up the green hue of the floor, while the wooden structure matches the wooden wall paneling.

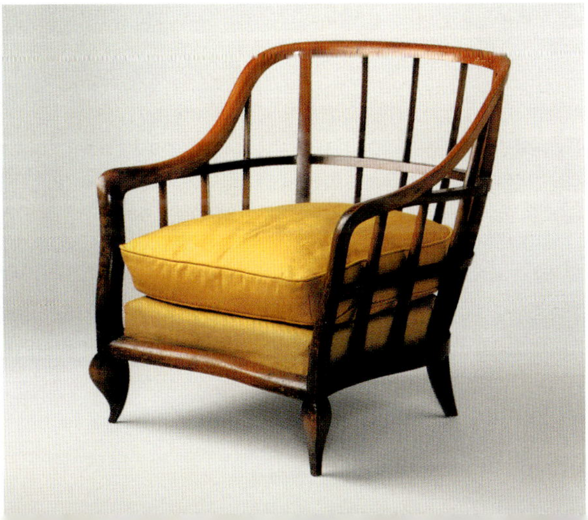

1937
armchair
designed for Alberto Pirelli
Private Collection

Luigi
Vietti

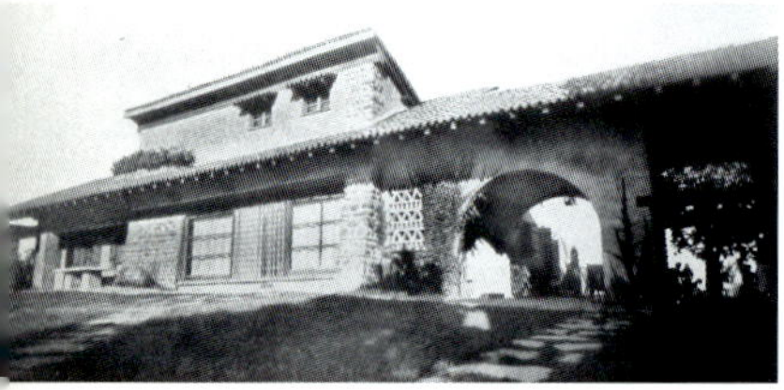

The hunting lodge at Cressa, the period facade and interiors, in Lo Stile nella casa e nell'arredamento, *1941.*

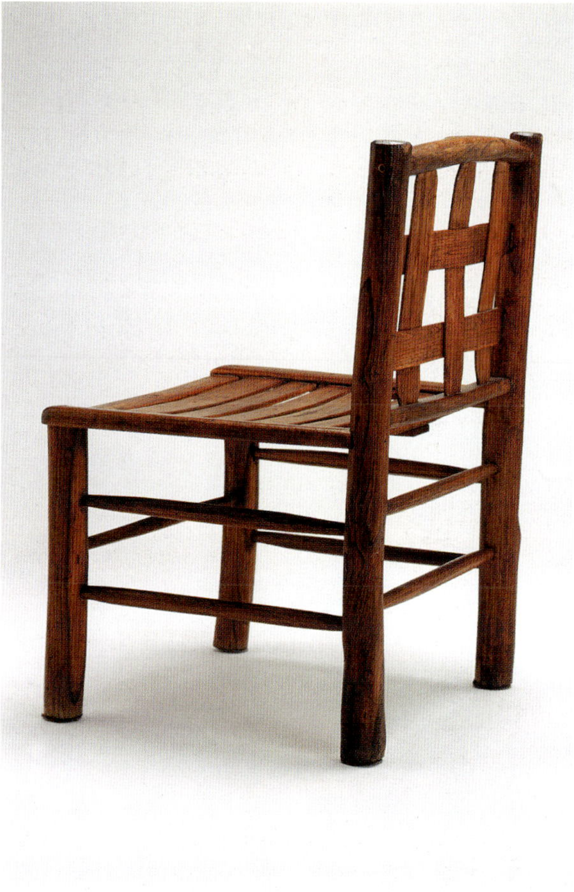

1939
chair
for the hunting lodge at Cressa, for Piero Galtrucco
Renato Veronesi Collection

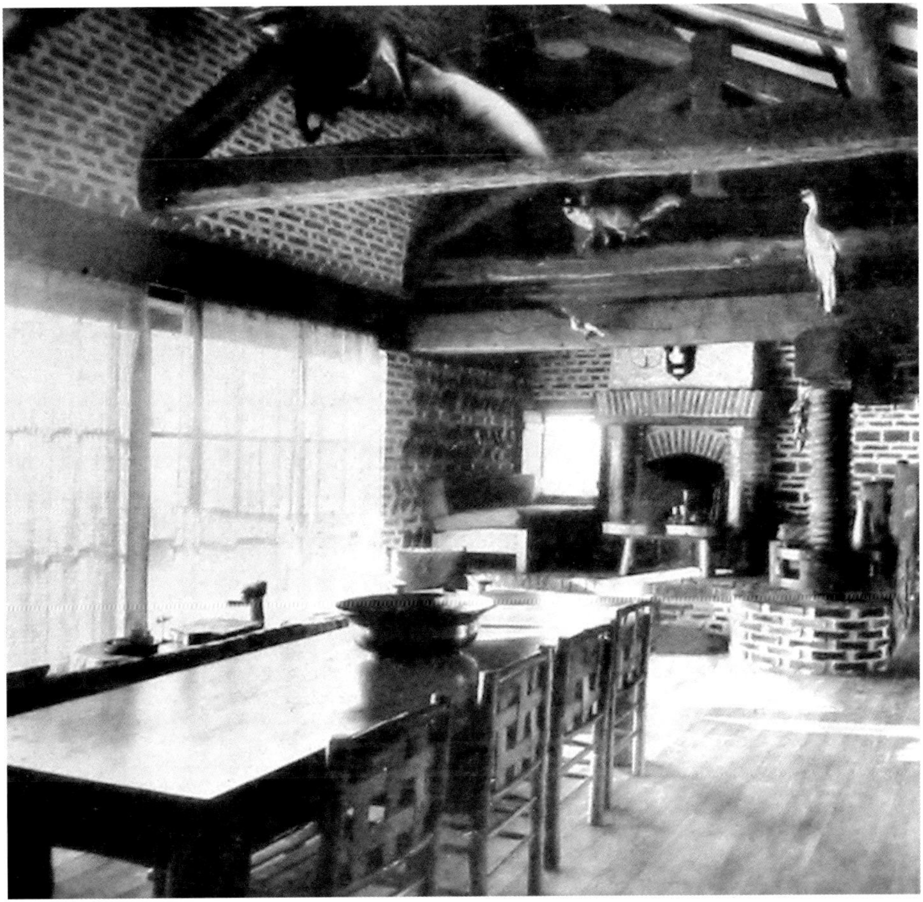

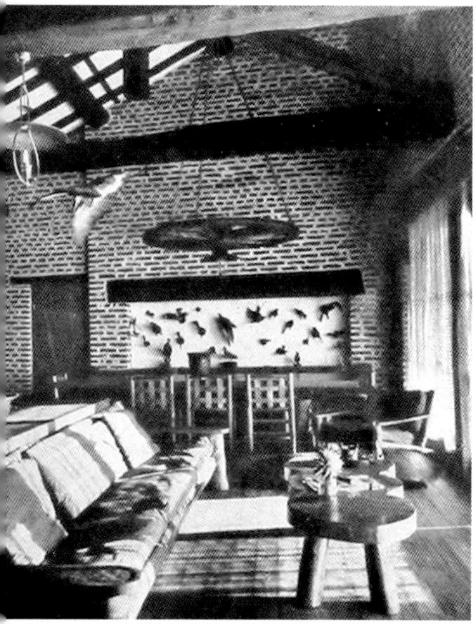

Designed for the hunting lodge at Cressa, located in the middle of an important reserve on a hill near Lake Maggiore, these two items represent a symbolic fusion of the archaic and modern, and craft and industry. Luigi Vietta always worked in this manner, using organic materials combined with new elements or treated with techniques refined by industry, with the intention of salvaging, updating and maintaining traditions and popular and vernacular tastes.
The armchair has a natural chestnut wooden frame, on which tow belts are laid to hold the cushions, while the workmanship of the seat – designed for the dining room – is ornate, the backrest made of woodskin woven into a lattice.
Used primarily for "vacation homes," for houses at the seaside, by the lake or in the mountains, where regard for nature and empathy with the landscape exert a powerful influence, Vietti's "method" expressed – in the Italy of the thirties and early forties – both the attraction and the resistance to any claim of the item of furniture as *product*: "I designed these chairs in the thirties. They wanted them in order to manufacture them, but I was possessive; three or four models are enough." (Interview given by Luigi Vietti to Rosanna Monzini in May 1991, see *Abitare*, no. 302, (December 1991), p. 59.

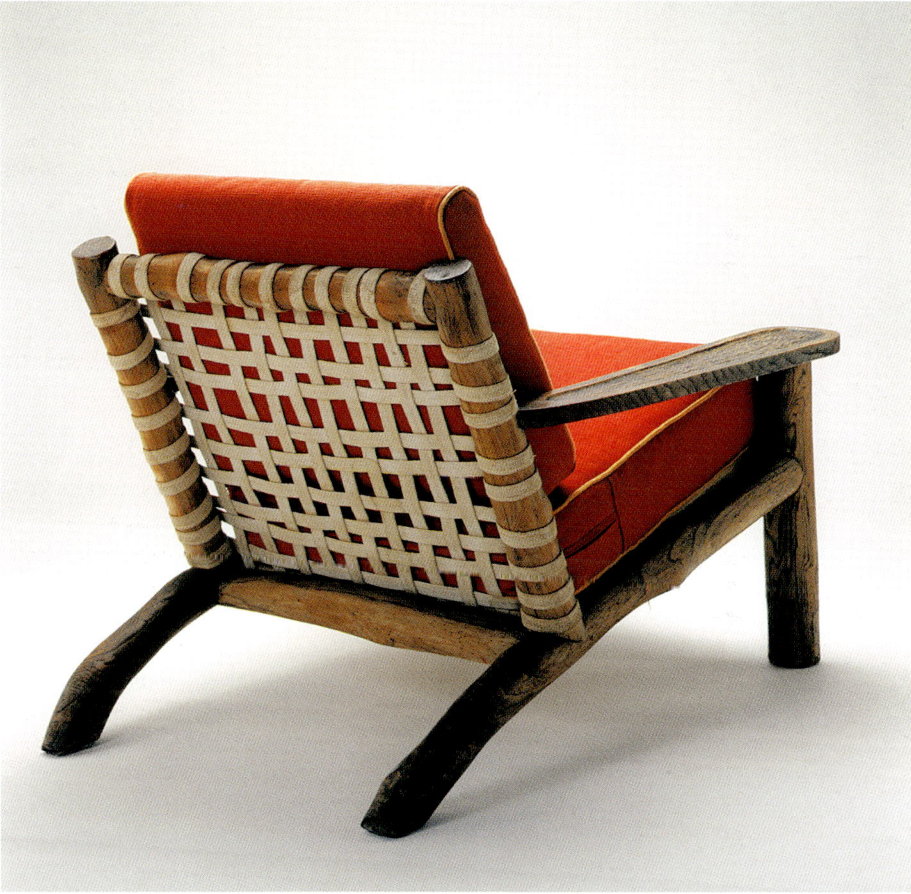

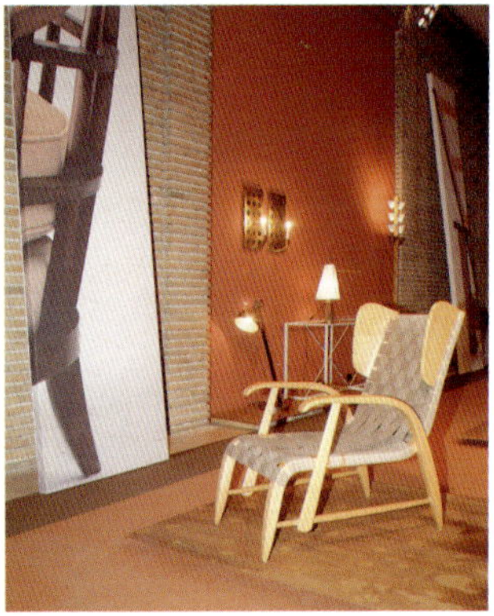

Installation

Giuseppe Raboni and Marco Zanuso

The exhibition installation takes into account its container, the Palazzo dell'Arte, a contemporary structure of the furniture on exhibit. Above all we attempted to emphasize the architecture of the *Impluvium,* making the most of the large central skylight and brick side walls.

Moreover, upon entering we wanted visitors to see all the furniture at once: to thus perceive the variety and complexity of the solutions adopted. We utilized long, narrow raised platforms, placed to create a route so that the public could approach and see the furniture easily from the sides and back.

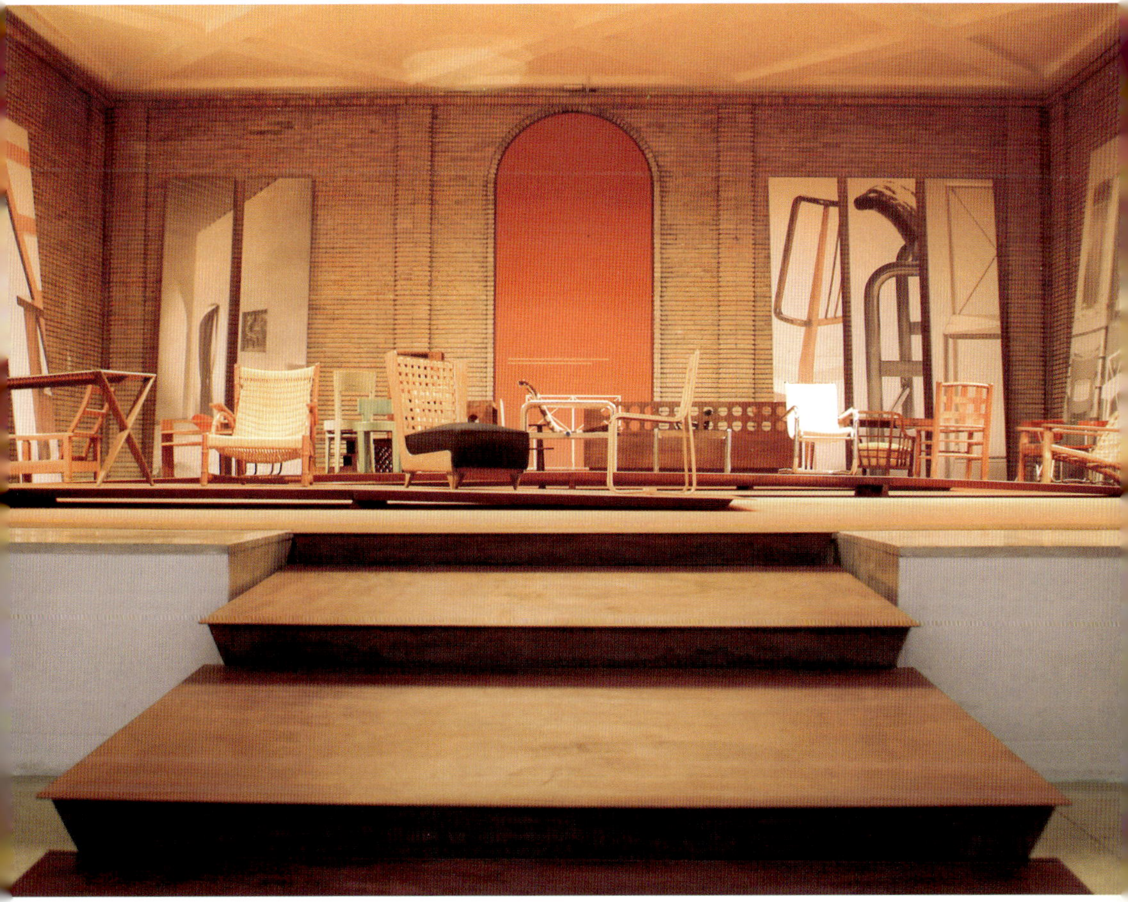

The raised platforms are made of folded and tarnished sheet metal, finished with a burnished brown wax and slightly curved to create something of a distorted perspective that contributes to a perception of the volumes of the exhibited objects.
The floor is covered with a large brick red "carpet" that gives the exhibition a warm and cozy character.
Panels lean against the walls, showing black and white archive photographs of the original environments, while detailed blowups highlight important features.

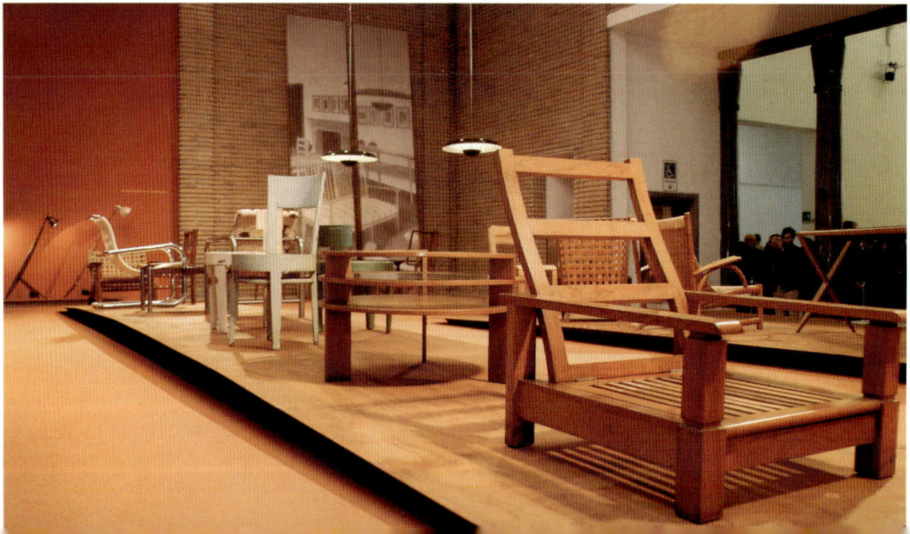

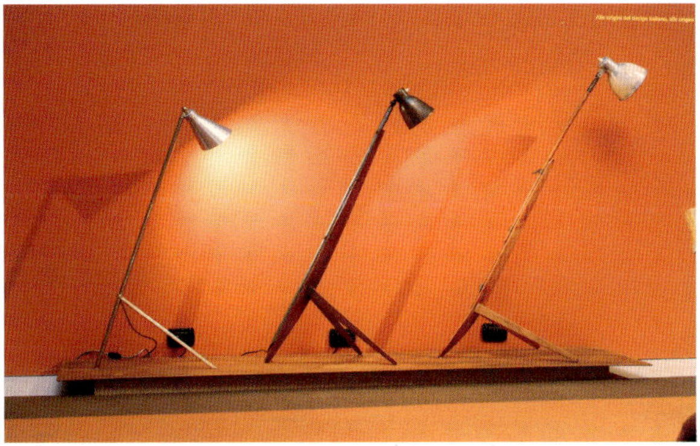

The light project was conceived in relation to the existing skylight: twelve shielded projectors with a diffuser panel measuring 4 x 4 meters were mounted inside it so that the characteristic effect of daylight from above would remain the same during the hours of darkness. Six groups of swiveling spotlights placed on brick pilasters complete the pattern.

We would like to thank the following
Archivio Asnago-Vender architetti · Archivio Fondazione Piero Portaluppi · Daniele Baroni · Archivio Piero Bottoni · Dpa, Politecnico di Milano · Marco Albini · Daniele Baroni · Alberto Bassi · Mirella e Vittorio Bega · Luigi Caccia Dominioni · Cristina Cappalegora · Laura Caronni · Elena Chiesa · Piergiovanni Ceregioli · Edoarda De Ponti · Monica Fried · Famiglia Gardella · Adolfo Guzzini · Monica Lattuada · Elena Lingeri · Francesca Luchi · Ferruccio Luppi · Giorgio Manusardi · Paolo Benedetto Manusardi · Mariangela Marelli · Paolo Minerbi · Tinoune Moise · Rosanna Monzini · Gianni Ottolini · Alessandro Pedretti · Roberto Poggi · Roberto Rizzi · Stefano Ronchetti · Marzio Rusconi Clerici · Valerio Terraroli · Giancorrado Ulrich · Renato Veronesi · Marcello Zocca

For their kind collaboration
Matteo Dolcetti · Enrico Mastrapasqua · Paolo Pelanda

Photographic Credits

Photographic Archives of La Triennale di Milano, photo Václav Šedý: pp. 36, 37, 40 above, 41, 43 at right, 45, 46, 48 at right, 49 at left, 51, 53 above, 54 at right, 57, 59, 61, 63, 65, 67, 68, 69 at right, 70, 71, 73, 74 at left, 75, 76, 78, 83, 84, 87 at left, 88, 89, 91, 92 at right, 93, 94, 98, 100, 101, 102 at right, 105

Historic Photographic Archives of La Triennale di Milano: pp. 39, 55, 74

Franco Albini Archives: p. 38
Vittorio Bega Archives: pp. 44, 48 at left, 49 at right
Pier Bottoni Archives, DPA, Politecnico di Milano: pp. 50, 52, 53 below, 54 at left
Cassina Archives, Meda: p. 95 photo Villani, Bologna, p. 96 photo E. Binelli, Genoa, p. 97
Piero Portaluppi Foundation Archives, Milan, Photo Antonio Paoletti: pp. 90, 92
Lorenzo Galtrucco S.p.A. Archives: p. 99
Studio Monzini-Raboni Archives: pp. 56, 58, 60

pp. 40, 42: from Airoldi Renato, *Asnago-Vender Architetti* (Como: Tipografia Editrice Cesare Nani, 1986)
p. 43: from *Casabella Costruzioni*, no. 146 (1940)
p. 47: from *Domus*, no. 165 (September 1941)
p. 66: from *Domus* 15, no. 109 (January 1937)
p. 69 at left: from *Rassegna*, no. 31 (1987)
p. 72: from *Domus*, no. 207 (March 1946)
pp. 79, 81: from *Domus* 20, no. 170 (February 1942)
pp. 82, 85, 86, 87 at right: from *Casabella Costruzioni*, no. 138-139-140 (June-August 1939)
pp. 102 at left, 103, 104: from *Lo stile nella casa e nell'arredamento* (1941)

Photographs of the installation: ag. 3effe_m.pandullo

To find out more about Charta, and to learn about our most recent publications, visit

www.chartaartbooks.it

Printed in May 2003
by La Grafica-Cantù, e-mail: lagrafica@cracantu.it
for Edizioni Charta